Ebola

Ebola

The Natural and Human History of a Deadly Virus

DAVID QUAMMEN

W. W. NORTON & COMPANY

New York • London

Ebola

INTRODUCTION

During the spring and summer of 2014, people around the world have watched with concern, appalled fascination, sympathy, and no small amount of personal fear as an outbreak of Ebola virus disease (EVD) has unfolded and spread among three troubled countries in West Africa—Guinea, Liberia, Sierra Leone—and then made a disconcerting leap, by airplane, to Nigeria. Having smoldered for months, killing victims by the dozens, it flamed in August of that year and began stacking up mortalities, week by week, in the hundreds. By then it had become the worst Ebola outbreak in the history of this peculiar, disconcerting disease. The story of the 2014 outbreak was so rivetingly awful that it competed for headline space with contemporaneous events in Syria, Ukraine, and the Gaza Strip.

But an outbreak of Ebola is very different from the dire realities of politics and war: more ineffable, more spooky. Ebola virus is invisible, except through an electron microscope or by way of its pathogenic effects. It is impersonal. It is apolitical. It seems to kill like the tenth plague of Egypt in Exodus—the one inflicted by an angel of death.

This last impression is misleading. Ebola is no death angel; it's mystifying but not preternatural. It's just a virus—albeit a virus that, inconspicuous elsewhere, tends to be hellaciously destructive when it gets into a human body.

Every newly emerging infectious disease, EVD included, begins as a mystery story. The mysteries are several. What's causing the sudden explosion of misery and death? If it's a virus, what sort of virus? Has science ever seen anything like it? Where has it come from? Any virus must abide in a living creature, in order to replicate and survive over time, so . . . which creature? And how has it moved from that creature into humans? Can the new virus be controlled? Can it be battled with pharmaceutical therapies or vaccines? Can it be stopped? Or is this outbreak going to be the Next Big One, a catastrophic pandemic, destined to sweep around the world and kill some sizable fraction of the human population, like the Black Death of the fourteenth century or the influenza of 1918? Disease scientists and public health officials are the intrepid investigators, the Sam Spades and Philip Marlowes and Detective Chief Inspector Jane Tennisons, who muster out to address these mysteries. In the case of Ebola, they have solved some but not all.

This little volume, excerpted and adapted from my 2012 book *Spillover*, with some additional material, is an attempt to place the 2014 West Africa outbreak—and a separate independent outbreak, which has recently flared in the Democratic Republic of the Congo—within a broader context that makes sense of those mysteries and their partial solutions. My offering here is merely a partial view of the history and science of Ebola, and a somewhat personal one, which has grown from my own modest travels through Ebola habitat, and from a chance encounter in the forest with two men who had seen the virus at its worst, killing their friends and loved ones. (To be clear: I myself have never had that harrowingly instructive experience, and I have not visited West Africa to observe or report on the current outbreak.) I also include here some treatment of Marburg virus, for two reasons: because it is closely related to Ebola virus, within the filovirus family, and because certain important questions that remain unanswered about Ebola virus *have* been answered for Marburg, as you'll see, suggesting valuable (though guarded) inferences about Ebola itself.

Ebola virus disease has been mostly an African affliction (so far), and although it's unique, it is no anomaly. It just represents an especially dramatic version of a global phenomenon.

Everything comes from somewhere, and strange new infectious diseases, emerging abruptly among humans, come mostly from nonhuman animals. The disease might be caused by a virus, or a bacterium, or a protozoan, or some other form of dangerous bug. That bug might live inconspicuously in a kind of rodent, or a bat, or a bird, or a monkey, or an ape. Crossing by some accident from its animal hideaway into its first human victim, it might find hospitable conditions; it might replicate aggressively and abundantly; it might cause illness, even death; and in the meantime, it might pass onward from its first human victim into others. There's a fancy word for this phenomenon, used by scientists who study infectious diseases from an ecological perspective: *zoonosis.*

That's a mildly technical term, unfamiliar to most people, but it helps clarify the biological complexities of swine flu, bird flu, SARS, West Nile fever, emerging diseases in general, and the threat of a global pandemic. It helps us comprehend why medical science and public health campaigns have been able to conquer some fearsome diseases, such as smallpox and polio, but are unable to conquer others, such as dengue and yellow fever. It's a word of the future, destined for heavy use in the twenty-first century. A zoonosis is an animal infection that's transmissible to humans.

Bubonic plague is a zoonosis. All strains of influenza are zoonoses. So are monkeypox, bovine tuberculosis, Lyme disease, Marburg, rabies, Hantavirus Pulmonary Syndrome, and a strange affliction called Nipah, which has killed pigs and pig farmers in Malaysia, as well as people who drink date palm sap (sometimes contaminated with the virus from bat droppings) in Bangladesh. Each of them reflects the action of a pathogen that can cross into people from other species. This form of interspecies leap is common, not rare; about 60 percent of all infectious diseases currently known either cross routinely or have recently crossed between other animals and us. Some of those—notably rabies—are familiar, widespread, and still horrendously lethal, killing humans by the thousands despite centuries of efforts at coping with their effects, concerted international attempts to eradicate or control them, and a pretty clear scientific understanding of how they work. Others are new and inexplicably sporadic, claiming a few victims or a few hundred in this place or that, and then disappearing for years.

Smallpox, to take one counterexample, is not a zoonosis. It's caused by the variola virus, which under natural conditions infects only humans. That helps explain why a global campaign mounted by the World Health Organization (WHO) to eradicate smallpox was, as of 1980, successful. Smallpox could be eradicated because that virus, lacking the ability to reside and reproduce anywhere but in a human body (or a carefully watched lab animal), couldn't hide.

Zoonotic pathogens can hide. That's what makes them so interesting, so complicated, and so problematic. These pathogens aren't *consciously* hiding, of course. They reside where they do and transmit as they do because those happenstance options have worked for them in the past, yielding opportunities for survival and reproduction. By the cold Darwinian logic of natural selection, evolution codifies happenstance into strategy.

The least conspicuous strategy of all is to lurk within what's called a reservoir host. A reservoir host is a species that carries the pathogen, harbors it chronically, while suffering little or no illness. When a disease seems to disappear between outbreaks, its causative agent has got to be *someplace*, yes? Well, maybe it vanished entirely from planet Earth—but probably not. Maybe it died off throughout the region and will only reappear when the winds and the fates bring it back from elsewhere. Or maybe it's still lingering nearby, all around, within some reservoir host. A rodent? A bird? A butterfly? A bat? To reside undetected within a reservoir host is probably easiest wherever biological diversity is high and the ecosystem is relatively undisturbed. The converse is also true: Ecological disturbance causes diseases to emerge. Shake a tree, and things fall out. Capture a bat for food, and you might catch something else too. Butcher a chimpanzee, to feed your family or your village, and who knows what grisly surprises might emerge. The event of transmission, when a pathogen passes from one kind of host to another, is called spillover.

Now you're equipped with the basic concepts. Here's the starting point for all that follows: Ebola is a zoonosis.

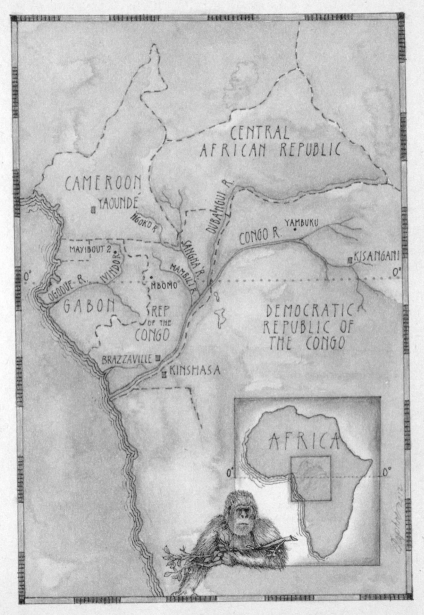

© Daphne Gillam

1

Along the upper Ivindo River in northeastern Gabon, near the border with the Republic of the Congo, lies a small village called Mayibout 2, a sort of satellite settlement, just a mile upriver from its namesake, the village of Mayibout. In early February 1996, this secondary community was struck by a horrific and bewildering chain of events. Eighteen people in Mayibout 2 became suddenly sick after they participated in the butchering and eating of a chimpanzee.

Their symptoms included fever, headache, vomiting, bloodshot eyes, bleeding from the gums, hiccupping, muscle pain, sore throat, and bloody diarrhea. All eighteen were evacuated downriver to a hospital in the district capital, a town called Makokou, by decision of the village chief. It's less than fifty miles as the crow flies from Mayibout 2 to Makokou, but by pirogue on the sinuous Ivindo, a journey of seven hours. The boats bearing victims wound back and forth between walls of forest along the banks. Four of the evacuees were moribund when they arrived and dead within two days. The four bodies, returned to Mayibout 2, were buried according to traditional ceremonial practice, with no special precautions against the transmission of whatever had killed them. A fifth victim escaped from the hospital, straggled back to the village, and died there. Secondary cases soon broke out among people infected while caring for the first victims—their loved ones or friends—or in handling the dead bodies. Eventually thirty-one people got sick, of whom twenty-one died: a case fatality rate of almost 68 percent.

Those facts and numbers were collected by a team of medical researchers, some Gabonese, some French, who reached Mayibout 2 during the outbreak. Among them was an energetic Frenchman named Eric M. Leroy, a Paris-trained virologist and veterinarian then based at the Centre International de Recherches Médicales de Franceville (CIRMF), in Franceville, a modest city in southeastern Gabon. Leroy and his colleagues identified the disease as Ebola hemorrhagic fever (a name now replaced by Ebola virus disease, reflecting the recognition that bloodiness is not quintessential) and deduced that the butchered chimpanzee had been infected with Ebola. "The chimpanzee seems to have been the index case for infecting 18 primary human cases," they wrote. Their investigation also turned up the fact that the chimp hadn't been killed by village hunters; it had been found dead in the forest and scavenged.

This was one piece of evidence, with much more to follow, that chimps and gorillas, like humans, are highly susceptible to Ebola. And if they suffer misery and speedy death from the virus, Leroy and other researchers reasoned, then they cannot be its reservoir host, the creature in which it abides inconspicuously over the long term. Instead, the dead chimp at Mayibout 2 was a clue. This sort of occasional role of an ape as an intermediate victim, catching the virus, passing it to humans, could perhaps help lead toward identification of the reservoir host itself. Was it some animal, large or small, with which chimpanzees come into contact?

Four years later, I sat at a campfire in deep forest near the upper Ivindo River, about forty miles due west of Mayibout 2. I was sharing dinner, from a big pot, with a dozen local men who were working as forest crew for a long overland trek. These men, most of them from villages in northeastern Gabon, had been walking for weeks before I joined them on the march. Their job involved carrying heavy bags through the jungle and building a simple camp each night for the biologist, one J. Michael Fay, whose obsessive sense of mission drove the whole enterprise forward. Mike Fay is an unusual man, even by the standards of tropical field biologists: physically tough, obdurate, free-spirited, smart, and fiercely committed to conservation. His enterprise, which he labeled the Megatransect, was a two-thousand-mile biological survey, on foot, through the wildest remaining forest areas of

Central Africa. He took data every step of the way, recording elephant dung piles and leopard tracks and chimpanzee sightings and botanical identifications, tiny notations by the thousands, all going into his waterproof yellow notebooks in scratchy left-handed print, while the crewmen strung out behind him toted his computers, his satellite phone, his special instruments and extra batteries, as well as tents and food and medical supplies enough for both him and themselves.

Fay had already been walking for 290 days by the time he reached this part of northeastern Gabon. He had crossed the Republic of the Congo with a field crew of forest-tough men, mostly Bambendjellés (one ethnic group of the short-statured peoples sometimes termed Pygmies), but those fellows had been disallowed entry at the Gabonese border. So Fay had been forced to raise a new team in Gabon. He recruited them largely from a cluster of gold-mining camps along the upper Ivindo River. The hard, stumbling work he demanded, cutting trail, schlepping bags, was evidently preferable to digging for gold in equatorial mud. One man served as cook as well as porter, stirring up massive amounts of rice or *fufu* (a starchy staple made from manioc flour, like an edible wallpaper paste) at each evening's campfire, and adorning it with some sort of indeterminate brown sauce. The ingredients for that variously included tomato sauce, dried fish, canned sardines, peanut butter, freeze-dried beef, and *pili-pili* (hot pepper), all deemed mutually compatible and combined at the whim of the chef. No one complained. Everyone was always hungry. The only thing worse than a big portion of such stuff, at the end of an exhausting day of stumbling through the jungle, was a small portion. My role amid this gang, on assignment for *National Geographic*, was to walk in Fay's footsteps and produce a series of stories describing the work and the journey. I would accompany him for ten days here, two weeks there, and then escape back to the United States, let my feet heal (we wore river sandals), and write an installment.

Each time I rejoined Fay and his team, there was a different logistical arrangement for our rendezvous, depending on the remoteness of his location and the urgency of his need to be resupplied. He never diverted from the zigzag line of his march. It was up to me to get to him. Sometimes I went in by bush plane and motorized dugout, along with Fay's trusted logistics man and quartermaster, a Japanese ecologist named Tomo Nishihara. Tomo and I would pile ourselves into the canoe amid whatever stuff he was bringing for the next leg of Fay's trek: fresh bags of fufu and rice and dried fish, crates of sardines, oil and peanut butter and pili-pili and double-A batteries. But even a dugout canoe couldn't always reach the spot where Fay and his crew, famished and bedraggled, would be waiting. On this occasion, with the trekkers crossing a big forest block called Minkébé, Tomo and I roared out of the sky in a Bell 412 helicopter, a massive 13-seater, chartered expensively from the Gabonese army. The forest canopy, elsewhere thick and unbroken, was punctuated here by several large granite gumdrops that rose above everything, hundreds of feet high, like El Capitan standing out of a green ground fog. Atop one of those inselbergs was the landing zone to which Fay had directed us. It was the only place within miles where a chopper could put down.

That day had been a relatively easy one for the crew—no swamps crossed, no thickets of skin-slicing vegetation, no charging elephants provoked by Fay's desire to take video at close range. They were bivouacked, awaiting the helicopter. Now the supplies had arrived—including even some beer! This allowed for a relaxed, genial atmosphere around the campfire. Quickly I learned that two of the crewmen, Thony M'Both and Sophiano Etouck, had roots in Mayibout 2, this famously unfortunate village about which I had read. They had been present when Ebola virus struck the village.

Thony, an extrovert, slim in build and far more voluble than the other fellow, was willing to talk about it. He spoke in French while Sophiano, a shy man with a body-builder's physique, an earnest scowl, a goatee, and a nervous stutter, sat silent. Sophiano, by Thony's account, had watched his brother and most of his brother's family die.

Having just met these two men, I couldn't decently press for more information that evening. Two days later we set off on the next leg of Fay's hike, across the Minkébé forest, heading southward away from the inselbergs. We got busy and distracted with the physical challenges of foot travel through trackless jungle terrain, and were exhausted (especially they, working harder than I) by nightfall. Halfway along, though, after a week of difficult walking, common miseries, and shared meals, Thony loosened enough to tell me more. His memories agreed generally with the report of the CIRMF team from Franceville, apart from small differences on some numbers and details. But his perspective was more personal.

Thony called it *l'épidémie*, the epidemic. This happened in 1996, yes, he said, around the same time some French soldiers came up to Mayibout 2 in a Zodiac raft and camped near the village. It was unclear whether the soldiers had a serious purpose—rebuilding an old airstrip?—or were just there to amuse themselves. They shot off their rifles. Maybe, Thony guessed, they also possessed some sort of chemical weaponry. He mentioned these details because he thought they might have relevance to the epidemic. One day some boys from the village went out hunting with their dogs. The intended prey was porcupines. Instead of porcupines they got a chimp—not killed by the dogs, no. A chimp found dead. They brought it back. The chimp was rotten, Thony said, its stomach putrid and swollen. Never mind, people were glad and eager for meat. They butchered the chimp and ate it. Then quickly, within two days, everyone who had touched the meat started getting sick.

They vomited; they suffered diarrhea. Some went downriver by motorboat to the hospital at Makokou. But there wasn't enough fuel to transport every sick person. Too many victims, not enough boats. Eleven people died at Makokou. Another eighteen died in the village. The special doctors quickly came up from Franceville, yes, Thony said, wearing their white suits and helmets, but they didn't save anyone. Sophiano lost six family members. One of those, one of his nieces—he was holding her as she died. Yet Sophiano himself never got sick. No, nor did I, said Thony. The cause of the illnesses was a matter of uncertainty and dark rumor. Thony suspected that the French soldiers, with their chemical weapons, had killed the chimpanzee and carelessly left its meat to poison the villagers. Anyway, his fellow survivors had learned their lesson. To this day, he said, no one in Mayibout 2 eats chimpanzee.

I asked about the boys who went hunting. Them, all the boys, they died, Thony said. The dogs did not die. Had he ever before seen such a disease, such an epidemic? *"Non,"* Thony answered. *"C'etait le premier fois."* Never.

How did they cook the chimp? I pried. In a normal African sauce, Thony said, as though that were a silly question. I imagined chimpanzee hocks in a peanutty gravy, with pili-pili, ladled over fufu.

Apart from the chimpanzee stew, one other stark detail lingered in my mind. It was something Thony had mentioned during our earlier conversation. Amid the chaos and horror in the village, Thony told me, he and Sophiano had seen something bizarre: a pile of thirteen gorillas, all dead, lying nearby in the forest.

Thirteen gorillas? I hadn't asked about dead wildlife. This was volunteered information. Of course, anecdotal testimony tends to be shimmery, inexact, sometimes utterly false, even when it comes from eyewitnesses. To say *thirteen dead gorillas* might actually mean a dozen, or fifteen, or simply lots—too many for an anguished brain to count. People were dying. Memories blur. To say *I saw them* might mean exactly that or possibly less. *My friend saw them, he's a close friend, I trust him like I trust my eyes.* Or maybe: *I heard about it on pretty good authority.* Thony's testimony, it seemed to me, belonged in the first epistemological category: reliable if not necessarily precise. I believed he saw these dead gorillas, roughly thirteen, in a group if not a pile; he may even have counted them. The image of thirteen gorilla carcasses strewn on the leaf litter was lurid but plausible. Subsequent evidence indicates that gorillas are highly susceptible to Ebola.

Scientific data are another matter, very different from anecdotal testimony. Scientific data don't shimmer with poetic hyperbole and ambivalence. They are particulate, quantifiable, firm. Fastidiously gathered, rigorously sorted, they can reveal emergent meanings. That's why Mike Fay was walking across Central Africa with his yellow notebooks: to search for big patterns that might emerge from masses of small data.

The next day we continued on through the forest. We were still more than a week from the nearest road. It was excellent gorilla habitat, well structured, rich with their favorite plant foods, and nearly untouched by humans: no trails, no camps, no evidence of hunters. It should have been full of gorillas. And once, in the recent past, it had been: A census of Gabon's ape populations done two decades earlier, by a pair of scientists from CIRMF, had yielded an estimate of 4,171 gorillas within the Minkébé forest bloc. Nevertheless, during our weeks of bushwhacking, we saw none. There was an odd absence of gorillas and gorilla sign—so odd that, for Fay, it seemed dramatic. This was exactly the sort of pattern, positive or negative, that his methodology was meant to illuminate. During the course of his entire Megatransect he recorded in his notebook every gorilla nest he saw, every mound of gorilla dung, every stem fed upon by gorilla teeth—as well as elephant dung, leopard tracks, and similar traces of other animals. At the end of our Minkébé leg, he subtotaled his data. This took him hours, holed away in his tent, collating the latest harvest of observations on his laptop. Then he emerged.

Over the past fourteen days, Fay informed me, we had stepped across 997 piles of elephant dung and not one dollop from a gorilla. We had passed amid millions of stems of big herbaceous plants, including some kinds (belonging to the family *Marantaceae*) with nutritious pith that gorillas devour like celery; but not one of those stems, so far as he'd noticed, had shown gorilla tooth marks. We had heard zero gorilla chest-beat displays, seen zero gorilla nests. It was like the curious incident of the dog in the nighttime—a silent pooch, speaking eloquently to Sherlock Holmes with negative evidence that something wasn't right. Minkébé's gorillas, once abundant, had disappeared. The inescapable inference was that something had killed them off.

2

The spillover at Mayibout 2 was no isolated event. It was part of a pattern of Ebola virus disease outbreaks across Central Africa—a pattern of which the meaning is still a matter of puzzlement and debate. The pattern stretches from 1976 (the first recorded emergence of Ebola virus) to the present in 2014, and from one side of the continent (Guinea, Liberia, Sierra Leone) to the other (Sudan and Uganda). The four major lineages of virus that have shown themselves during those emergence events are collectively known as ebolaviruses. On a smaller scale, within Gabon alone, there has been a tight clustering of Ebola incidents: three in less than two years, and all three rather closely localized in space. Mayibout 2 was the middle episode of that cluster.

An earlier outbreak began during December 1994 in the gold-mining camps on the upper Ivindo, the same area from which Mike Fay later recruited his Gabonese crew. These camps lie about twenty-five miles upstream from Mayibout 2. At least thirty-two people got sick, showing the usual range of symptoms (fever, headache, vomiting, diarrhea, and some bleeding) that suggest Ebola virus disease. The source was hard to pinpoint, though one patient told of having killed a chimpanzee that had wandered into his camp and acted strangely. Maybe that animal was infected, bringing the contagion to hungry humans. According to another account, the first case was a man who had come across a dead gorilla, took parts of it back to his camp, and shared. He died and so did others who touched the meat. Around the same time came some reports of chimps, as well as gorillas, seen dead in the forest. More generally, the miners (and their families—these camps were essentially villages), by their very presence, their needs for food, shelter, and fuel, had caused disturbance to the forest canopy and the creatures that lived in it.

From the mining camps, those victims in 1994 were transferred downriver (as they would be again from Mayibout 2) to Makokou General Hospital. Then arose a wave of secondary cases, focused around the hospital or in villages nearby. In one of those villages was a *nganga*, a traditional healer, whose house may have been a point of transmission between a certain mining-camp victim of the outbreak, seeking folk medicine, and an unlucky local person visiting the healer about something less dire than Ebola. Possibly the virus was passed by the healer's own hands. Anyway, by the time this sequence ended, forty-nine cases had been diagnosed, with twenty-nine deaths, for a case fatality rate of almost 60 percent.

A year later came the outbreak at Mayibout 2, second in the series. Eight months after that, the CIRMF scientists and others responded to a third outbreak, this one near the town of Booué in central Gabon.

The Booué situation had probably begun three months earlier, in July 1996, with the death of a hunter at a timber camp known as SHM, about forty miles north of Booué. In retrospect, this hunter's fatal symptoms were recognized as matching Ebola virus disease, though his case hadn't triggered alarm at the time. Another hunter died mysteriously in the same logging camp six weeks later. Then a third. What sort of meat were they supplying to the camp? Probably a wide range of wild creatures, including monkeys, duikers, bush pigs, porcupines, possibly even (despite legal restrictions) apes. And again there were reports of chimpanzees seen dead in the forest— fallen dead, that is, not shot dead. The three early human cases seem to have been independent of one another, as though each hunter contracted the virus from the wild. Then the third hunter broadened the problem, making himself a transmitter as well as a victim.

He was hospitalized briefly at Booué but left that facility, eluded medical authorities, went to a nearby village, and sought help there from another nganga. Despite the healer's ministrations the hunter died—and then so did the nganga and the nganga's nephew. A cascade had begun. During October and into succeeding months there was a wider incidence of cases in and around Booué, suggesting more person-to-person transmission. Several patients were transferred to hospitals in Libreville, Gabon's capital, and died there. A Gabonese doctor, having performed a procedure on one of those patients, fell sick himself and, showing little confidence in his own country's health care, flew to Johannesburg for treatment. That doctor survived, but a South African nurse who looked after him sickened and died. Ebola virus had thereby emerged from Central Africa into the continent at large. The eventual tally from this third outbreak, encompassing Booué, Libreville, and Johannesburg, was sixty cases, of which forty-five were fatal. Case fatality rate? For that one, you can do the math in your head.

Amid this welter of cases and details, a few common factors stand out: forest disruption at the site of the outbreak, dead apes as well as dead humans, secondary cases linked to hospital exposure or traditional healers, and a high case fatality rate, ranging from 60 to 75 percent. Sixty percent is extremely high for any infectious disease (except rabies); it's probably higher, for instance, than fatalities from bubonic plague in medieval France at the worst moments of the Black Death.

In the years since 1996, other outbreaks of Ebola virus disease have struck both people and gorillas within the region surrounding Mayibout 2. One area hit hard lies along the Mambili River, just over the Gabon border in northwestern Congo, another zone of dense forest encompassing several villages, a national park, and a recently created reserve known as the Lossi Gorilla Sanctuary. Mike Fay and I had walked through that area also, in March 2000, just four months before my rendezvous with him at the Minkébé inselbergs. In stark contrast to the emptiness of Minkébé, gorillas had been abundant within the Mambili drainage when we saw it. But two years later, in 2002, a team of researchers at Lossi began finding gorilla carcasses, some of which tested positive for antibodies to Ebola virus. (A positive test for antibodies is less compelling evidence than a find of live virus, but still suggestive.) Within a few months, 90 percent of the individual gorillas they had been studying (130 of 143 animals) had vanished. How many had simply run away? How many were dead? Extrapolating rather loosely from confirmed deaths and disappearances to overall toll throughout their study area, the researchers published a paper in *Science* under the forceful (but overconfident) headline: EBOLA OUTBREAK KILLED 5000 GORILLAS.

3

I n 2006 I returned to the Mambili River, this time with a team led by William B. (Billy) Karesh, then director of the Field Veterinary Program for the Wildlife Conservation Society (WCS) and now filling a similar role at EcoHealth Alliance, an organization devoted to the study and prevention of zoonotic outbreaks all over the world. Billy Karesh is a veterinarian and an authority on zoonoses. He's a peripatetic field man, raised in Charleston, South Carolina, nourished on Marlin Perkins, whose usual working uniform is a blue scrub shirt, a gimme cap, and a beard. An empiricist by disposition, he speaks quietly, barely moving his mouth, and avoids categorical pronouncements as though they might hurt his teeth. Often he wears a sly smile, suggesting amusement at the wonders of the world and the varied spectacle of human folly. But there was nothing amusing about his mission to the Mambili. He had come to shoot gorillas—not with bullets but with tranquilizer darts. He meant to draw blood samples and test them for antibodies to Ebola virus.

Our destination was a site known as the Moba Bai complex, a group of natural clearings near the east bank of the upper Mambili, not far from the Lossi sanctuary. A *bai* in Francophone Africa is a marshy meadow, often featuring a salt lick, and surrounded by forest like a secret garden. In addition to Moba Bai, the namesake of this complex, there were three or four others nearby. Gorillas (and other wildlife) frequent such bais, which are waterlogged and sunny, because of the sodium-rich sedges and asters that grow beneath the open sky. We arrived at Moba, coming upstream on the Mambili, in an overloaded dugout pushed by a 40-horse outboard.

The boat carried eleven of us and a formidable pile of gear. We had a gas-powered refrigerator, two liquid-nitrogen freezer tanks (for preserving samples), carefully packaged syringes and needles and vials and instruments, examination gloves, hazmat suits, tents and tarps, rice, fufu, canned tuna, canned peas, several boxes of bad red wine, numerous bottles of water, a couple of folding tables, and seven stackable white plastic chairs. With these tools and luxurious provisions we established a field camp across the river from Moba. Our team included an expert tracker named Prosper Balo, plus other wildlife veterinarians, other forest guides, and a cook. Prosper had worked at Lossi before and during the outbreak. With his guidance, we would prowl the complex of bais, all full of succulent vegetation and previously famed for the dozens of gorillas that came there daily to eat and relax.

Billy Karesh had visited the same area twice previously, before Ebola struck, seeking baseline data on gorilla health. During a 1999 trip, he had seen sixty-two gorillas here in one day. In 2000 he returned to try darting a few. "Every day," he told me, "every bai had at least a family group." Not wanting to be too disruptive, he had tranquilized only four animals, weighed them and examined them for obvious diseases (such as yaws, a bacterial skin infection), and taken blood samples. All four apes had tested negative for Ebola antibodies. This time things were different. He wanted blood serum from survivors of the 2002 die-off. So we began, with high expectations. Days passed. As far as we could see, there *were* no survivors.

Precious few, anyway—not enough to make gorilla-darting (which is always a parlous enterprise, with some risk for both the darter and the dartees) productive of data. Our stakeout at Moba lasted more than a week. Early each morning we crossed the river, walked quietly to one bai or another, concealed ourselves in thick vegetation along the edge, and waited patiently for gorillas to appear. None did. Often we hunkered in the rain. When it was sunny, I read a thick book or dozed on the ground. Karesh stood ready with his air rifle, the darts loaded full of tilletamine and zolazepam, drugs of choice for tranquilizing a gorilla. Or else we hiked through the forest, following closely behind Prosper Balo as he searched for gorilla sign and found none.

On the morning of day 2, along a swampy trail to the bais, we saw leopard tracks, elephant tracks, buffalo tracks, and chimpanzee sign, but no evidence of gorillas. On day 3, with still no gorillas, Karesh said: "I think they're dead. Ebola went through here." He figured that only a lucky few, uninfected by the disease or else resistant enough to survive it, remained. Then again, he said, "those are the ones we're interested in," because they, if any, might carry antibodies. On day 4, separating from the rest of us, Karesh and Balo managed to locate a single, distraught male gorilla from the sound of his chest beats and screaming barks, and to crawl within ten yards of him in the thick underbrush. Suddenly the animal stood, only his head visible, in front of them. "I could have killed him," Karesh said later. "Pitted him." Drilled him between the eyes, that is, but not immobilized him with a safe shot to the flank. So Karesh held his fire. The gorilla let out another bark and ran off.

My notes from day 6 include the entry: "Nada nada nary gorilla nada." On our final chance, day 7, Balo and Karesh tracked another couple of animals for hours through the boggy forest without getting so much as a good glimpse. Gorillas had become desperately scarce, round about Moba Bai, and the stragglers were fearfully shy. Meanwhile the rain continued, the tents grew muddy, and the river rose.

When we weren't in the forest, I spent time in camp talking with Karesh and the three Africa-based WCS veterinarians on his team. One was Alain Ondzie, a lanky and bashful Congolese, trained in Cuba, fluent in Spanish as well as French and several Central African languages, with a likable tendency to dip his head and giggle joyously whenever he was teased or amused. Ondzie's main job was to respond to reports of dead chimps or gorillas anywhere in the country, getting to the site as quickly as possible and taking tissue samples to be tested for Ebola virus. He described to me the tools and procedures for such a task, with the carcass invariably putrefied by the time he reached it and the presumption (until otherwise proven) that it might be seething with Ebola. His working costume was a disposable hazmat suit with a vented hood, rubber boots, a splash apron, and three pairs of gloves, duct taped at the wrists. Making the first incision for sampling was dicey because the carcass might have become bloated with gas; it could explode. In any case the dead ape was usually covered with scavenging insects—ants, tiny flies, even bees. Ondzie told of one occasion when three bees from a carcass ran up his arms, under his hood flap, down across his bare body, and commenced to sting him as he worked on the samples. Can Ebola virus travel on the stinger of a bee? Probably, for at least a short period of time, if you're not lucky. Ondzie was lucky.

Does this work frighten you? I asked him. Not anymore, he said. Why do you do it? I asked. Why do you love it? (as he clearly did). *"Ça, c'est une bonne question,"* he said, with the characteristic bob and giggle. Then he added, more soberly: Because it allows me to apply what I've learned, and to keep learning, and it might save some lives.

Another member of the team was Patricia (Trish) Reed, who had come out to Africa as a biologist fifteen years earlier, studied Lassa fever and then AIDS, hired on with CIRMF in France-ville, gotten some field experience in Ethiopia, and then collected a DVM from the veterinary school at Tufts University in Boston. She was back at CIRMF, doing research on a monkey virus, when the WCS field vet working out here was killed in a plane crash coming into a backcountry Gabonese airstrip. Karesh hired Reed as the dead woman's replacement.

The scope of her work, Reed told me, encompassed a range of infectious diseases that threaten gorilla health, of which Ebola is only the most exotic. The others were largely human diseases of more conventional flavor, to which gorillas are susceptible because of their close genetic similarity to us: TB, poliomyelitis, measles, pneumonia, chickenpox, et cetera. Gorillas can be exposed to such infections wherever unhealthy people are walking, coughing, sneezing, and crapping in the forest. Any such spillover in the reverse direction—from humans to nonhuman animals—is known as an *anthroponosis*. The famous mountain gorillas, for instance, have been threatened by anthroponotic infections such as measles, carried by ecotourists who come to dote upon them. (Mountain gorillas constitute a severely endangered subspecies of the eastern gorilla, confined to the steep hillsides of the Virunga Volcanoes in Rwanda and neighboring lands. The western gorilla of Central African forests, a purely lowland species, is more numerous but far from secure.) Combined with destruction of their habitat by logging operations, and the hunting of them for bushmeat to be consumed locally or sold into markets, infectious diseases could push western gorillas from their current levels of relative abundance (maybe a hundred thousand in total) to a situation in which small, isolated populations survive tenuously, like the mountain gorillas, or go locally extinct.

But the forests of Central Africa are still relatively vast, compared to the small Virunga hillsides that harbor mountain gorillas; and the western gorilla doesn't face many ecotourists in its uncomfortable, nearly impenetrable home terrain. So measles and TB aren't the worst of its problems. "I would say that, without a doubt, Ebola is the biggest threat" to the western species, Reed said.

What makes Ebola virus among gorillas so difficult, she explained, is not just its ferocity but also the lack of data. "We don't know if it was here before. We don't know if they survive it. But we need to know how it passes through groups. We need to know *where* it *is*." And the question of *where* has two dimensions. How broadly is Ebola virus distributed across Central Africa? Within what reservoir host does it lurk?

On the eighth day, we packed up, reloaded the boats, and departed downstream on the Mambili, taking away no blood samples to add to the body of data. Our mission had been thwarted by the very factor that made it relevant: a notable absence of gorillas. Here was the curious incident of the dog in the nighttime again. Billy Karesh had seen one gorilla at close range but been unable to dart it, and had tracked two others with the help of Prosper Balo's keen eye for spoor. The rest, the many dozens that formerly frequented these bais, had either dispersed to parts unknown or they were . . . dead? Anyway, once gorillas had been abundant hereabouts, and now they were gone.

The virus seemed to be gone too. But we knew it was only hiding.

4

Hiding where? For almost four decades, the identity of Ebola's reservoir host has been one of the darkest little mysteries in the world of infectious disease. That mystery, along with efforts to solve it, dates back to the first recognized emergence of Ebola virus disease, in 1976.

Two outbreaks occurred in Africa that year, independently but almost simultaneously: one in the north of Zaire (now the Democratic Republic of the Congo) and one in south-western Sudan (in an area that today lies within the Republic of South Sudan), the two separated by three hundred miles. Although the Sudan situation began slightly earlier, the Zaire event is the more famous, partly because a small waterway there, the Ebola River, eventually gave its name to the virus.

The focal point of the Zaire outbreak was a small Catholic mission hospital in a village called Yambuku, within the district known as Bumba Zone. In mid-September, a Zairian doctor there reported two dozen cases of a dramatic new illness— not the usual malarial fevers but something more grisly, more red, characterized by bloody vomiting, nosebleeds, and bloody diarrhea. Fourteen of the patients had died, as of the doctor's cabled alert to authorities in Kinshasa, Zaire's capital, and others were in danger. By the start of October, Yambuku Mission Hospital had closed, for the grim reason that most of its staff members were dead. An international response team of scientists and physicians converged on the area several weeks later, under the direction of the Zairian Minister of Health, to do a crash study of the unknown disease and give advice toward controlling it. This group, consisting of members from France, Belgium, Canada, Zaire, South Africa, and the United States, including nine from the Center for Disease Control (later to become the Centers for Disease Control and Prevention, but then, and still, CDC) in Atlanta, became known as the International Commission. Their leader was Karl Johnson, an American physician and virologist who had gained high regard as well as experience from his earlier work on dangerous new pathogens, most notably one called Machupo virus, in Bolivia, back in 1963, which had eventually infected Johnson himself and nearly killed him. Thirteen years later, still intense, still dedicated, and not noticeably mellowed by near-death experience or professional ascent, he was head of the Special Pathogens Branch at the CDC.

Johnson had helped solve the Machupo crisis by his attention to the ecological dimension—that is, where did the virus live when it wasn't killing Bolivian villagers? The reservoir question had been tractable, in that case, and the answer had quickly been found: A native mouse was carrying Machupo into human households and granaries. Trapping out the mouse effectively ended the outbreak. Now, amid the desperate and befuddling days of October and November 1976, in northern Zaire, confronting a different invisible and unidentified killer, as the death toll rose into the hundreds, Johnson and his fellow researchers found time to wonder about Ebola virus as he had wondered about Machupo virus: Where did this thing come from?

By then they knew that the Zaire pathogen *was* a virus. That knowledge derived from isolations performed quickly on clinical samples shipped to overseas laboratories, including the CDC. (Johnson, before flying to Zaire, had led the CDC isolation effort himself.) They knew that this virus was similar to Marburg virus, another lethal agent, identified nine years before; the electron micrographs showed that it was equally filamentous and twisty, like an anguished tapeworm. But the lab tests also revealed Ebola virus as distinct enough from Marburg virus to constitute something new. Eventually these two wormy viruses, Ebola and Marburg, would be classified within a new family, *Filoviridae*: the filoviruses.

Johnson's group knew also that the new agent, Ebola virus, must reside in some living animal—something other than humans—where it could exist less disruptively and maintain a continuous presence. But the question of its reservoir was less urgent than other concerns, such as how to break the chain of person-to-person transmission, how to keep patients alive, how to end the outbreak. "Only limited ecological investigations were made," the team reported later, and the results of those investigations were all negative. No sign of Ebola virus appeared anywhere except in humans. But the negative data are interesting in retrospect, at least as a record of where these early researchers looked. They pureed 818 bedbugs collected from Ebola-affected villages, finding no evidence of the virus in any. They considered mosquitoes. Nothing. They drew blood from ten pigs and one cow—all of which proved Ebola-free. They caught 123 rodents, including 69 mice, 30 rats, and 8 squirrels, not one of which was a viral carrier. They read the entrails of six monkeys, two duikers, and seven bats. These animals also were clean.

The International Commission members were chastened by what they had seen. "No more dramatic or potentially explosive epidemic of a new acute viral disease has occurred in the world in the past 30 years," their report warned. The case fatality rate of 88 percent, they noted, was higher than any on record, apart from the rate for rabies (almost 100 percent among patients not treated before they show symptoms). The Commission made six urgent recommendations to Zairian officialdom, among which were health measures at the local level and nationwide surveillance. But the identification of Ebola's reservoir wasn't mentioned. That was a scientific matter, slightly more abstract than the action items offered to President Mobutu's government. It would have to wait.

The wait has continued.

Three years after Yambuku, Karl Johnson and several other members of the Commission were still wondering about the reservoir question. They decided to try again. Lacking funds to mount an expedition devoted solely to finding Ebola's hideout, they hitched their effort to an ongoing research program on monkeypox in Zaire, coordinated by the World Health Organization. Monkeypox is a severe affliction, though not so dramatic as Ebola virus disease, and also caused by a virus that lurks in a reservoir host or hosts, at that time still unidentified. So it seemed natural and economical to do a combined search, using two sets of analytical tools to screen a single harvest of specimens. Again the field team collected animals from villages and surrounding forest in Bumba Zone, as well as in other areas of northeastern Zaire and southeastern Cameroon. This time their trapping and hunting efforts, plus the bounties they paid for creatures delivered alive by villagers, yielded more than fifteen hundred animals representing 117 species. There were monkeys, rats, mice, bats, mongooses, squirrels, pangolins, shrews, porcupines, duikers, birds, tortoises, and snakes. Blood was taken from each, and then snips of liver, kidney, and spleen. All these samples, deep-frozen in individual vials, were shipped back to the CDC for analysis. Could any live virus be grown from the sampled tissues? Could any Ebola antibodies be detected in blood serum? The bottom line, reported with candor by Johnson and coauthors in the pages of *The Journal of Infectious Diseases*, was negatory: "No evidence of Ebola virus infection was found."

One factor making the hunt for Ebola's reservoir especially difficult, especially hard to focus, is the transitory nature of the disease within human populations. It disappears entirely for years at a time. This is a mercy for public health but a constraint for science. Viral ecologists can look for Ebola anywhere, in any creature of any species, in any African forest, but those are big haystacks and the viral needle is small. The most promising search targets, in space and in time, are wherever and whenever people are dying of Ebola virus disease. And for a long interlude, no one *was* dying of that disease—no one whose death came to the attention of medical authorities, anyway.

After the Yambuku outbreak of 1976, and then two episodes in Zaire and Sudan between 1977 and 1979, ebolaviruses barely showed themselves anywhere in Africa for fifteen years. There may have been some scattered cases during the early 1980s, retrospectively suspected, but there was no confirmed outbreak that evoked emergency response; and in each of those minor instances the chain of infection seemed to have burned itself out. Burning out is a concept with special relevance to such highly lethal and moderately contagious pathogens. It means that a few people died, a few more got infected, a fraction of those also died but others recovered, and the pathogen didn't continue to propagate. The incident expired on its own before shock troops from the WHO, the CDC, and other centers of expertise had to be mustered. Then, after an interval, it returned—with the outbreaks at Mayibout 2 and elsewhere in Gabon, and even more alarmingly at a place called Kikwit.

Kikwit, in Zaire, lay about three hundred miles east of Kinshasa. It differed from Yambuku, and Mayibout 2, and the timber camp outside Booué in one crucial way: It was a city of two hundred thousand people. It contained several hospitals. It was connected to the wider world in a way that those other outbreak sites weren't. But like them it was surrounded by forest.

The first identified case in the Kikwit outbreak was a forty-two-year-old man who worked in or near that forest and probably, to some small extent, disturbed it. He farmed several patches of cleared land, planting corn and cassava, and made charcoal from timber, all at a spot five miles southeast of the city. How did he get his wood supply, how did he clear daylight for his gardens? Presumably by cutting trees. This man fell sick on January 6, 1995, and died of a hemorrhagic fever a week later.

By that time he had directly infected at least three members of his family, all fatally, and launched the infection into his wider circle of social contacts, ten more of whom died within coming weeks. Some of those contacts evidently carried the virus into the city's maternity hospital, where it infected a laboratory technician, and from there into Kikwit General Hospital. The technician, while being treated at Kikwit General, infected several doctors and nurses who did surgery on him (suspecting a gut perforation related to typhoid, they cut open his abdomen), as well as two Italian nuns who helped with his care. The technician died, the nuns died, and local officials hypothesized that this was epidemic dysentery, a misdiagnosis that allowed the virus to spread further among patients and staff at other hospitals in the Kikwit region.

Not everyone accepted the dysentery hypothesis. One doctor at the Ministry of Health thought it looked instead like a viral hemorrhagic fever, which suggested Ebola. That good guess was confirmed quickly from blood specimens received by the CDC, in Atlanta, on May 9. Yes indeed: It was Ebola virus. By the end of the outbreak, in August, 245 people had died, including 60 hospital staff members. Performing abdominal surgery on Ebola patients, when you thought they were suffering from something else (such as gastrointestinal bleeding from ulcers), was risky work.

Meanwhile, another international team came out to search for the reservoir, converging on Kikwit in early June. This group consisted of people from the CDC, from a Zairian university, from the United States Army Medical Research Institute of Infectious Diseases (USAMRIID, formerly a bioweapons lab but now committed to disease research and biodefense) in Maryland, and one fellow from the Danish Pest Infestation Laboratory, who presumably knew a lot about rodents. They began work at the site to which the spillover seemed traceable—that is, at the charcoal pit and crop fields of the unlucky forty-two-year-old man, the first victim, southeast of the city. From that site and others, over the following three months, they trapped and netted thousands of animals. Mostly those were small mammals and birds, plus a few reptiles and amphibians. All the traps were set within forest or savanna areas outside the city limits. Within Kikwit itself, the team netted bats at a Sacred Heart mission. They killed each captured animal, drew blood, and dissected out the spleen (in some cases other organs too, such as a liver or a kidney), which went into frozen storage. They also took blood from some dogs, cows, and pet monkeys. The total yield included 3,066 blood samples and 2,730 spleens, all shipped back to the CDC for analysis. The blood samples, after having been irradiated to kill any virus, were tested for Ebola virus antibodies, using the best available molecular method of the time. The spleens were transferred to a biosafety level 4 (BSL-4) laboratory, a new sort of facility since Karl Johnson's early work (and of which he was one of the pioneering designers), with multiple seals, negative air pressure, elaborate filters, and lab personnel working in spacesuits—a containment zone in which Ebola virus could be handled without risk (theoretically) of accidental release. No one knew whether any of these Zairian spleens contained the virus but each had to be treated as though it did. From the spleen material, minced finely and added to cell cultures, the lab people tried to grow the virus.

None grew. The cell cultures remained blithely unspotted by viral blooms. And the antibody tests yielded no positive hits either. Once again, Ebola virus had spilled over, caused havoc, and then disappeared without showing itself anywhere but in the sick and dying human victims. It was Zorro, it was the Swamp Fox, it was Jack the Ripper—dangerous, invisible, gone.

This three-month, big-team effort at Kikwit shouldn't be considered a total failure; even negative results from a well-designed study tend to reduce the universe of possibilities. But it was another hard try ending in frustration. Maybe the Kikwit team had gotten there too late, five months after the charcoal maker fell ill. Maybe the shift from wet season to dry season had caused the reservoir, whatever it is, to migrate or hide or decrease in abundance. Maybe the virus itself had declined to a minimal population, a tenuous remnant, undetectable even within its reservoir during the off season. The Kikwit team couldn't say. The most notable aspect of their eventual report, apart from its long list of animals that *didn't* contain Ebola virus, was its clear statement of three key assumptions that had guided their search.

First, they suspected (based on earlier studies) that the reservoir is a mammal. Second, they noted that Ebola virus disease outbreaks in Africa had always been linked to forests. (Even the urban epidemic at Kikwit had begun with that charcoal maker out amid the woods.) It seemed safe to assume, therefore, that the reservoir is a forest creature. Third, they noted also that Ebola outbreaks had been sporadic in time—with years sometimes passing between one episode and the next. Those gaps implied that infection of humans from the reservoir is a rare occurrence. Rarity of spillover in turn suggested two possibilities: that either the reservoir itself is a rare animal or that it's an animal only rarely in contact with people.

Beyond that, the Kikwit team couldn't say. They published their paper in 1999 (among a whole series of reports on Ebola, in a special supplement of *The Journal of Infectious Diseases*), authoritatively documenting a negative conclusion. After twenty-three years, the reservoir still hadn't been found.

5

"We need to know where it is," Trish Reed had said. She was alluding to the two unanswered questions about Ebola virus and its location in space. The first question is ecological: In what living creature does it hide? That's the matter of reservoir. The second question is geographical: What's its distribution across the African landscape? The second may be impossible to answer until the reservoir is identified and *its* distribution traced. In the meantime, the only data reflecting Ebola virus's whereabouts are the plotted points of human outbreaks on a map.

Let's glance across that map. In 1976 Ebola virus made its debut, as I've mentioned, with the dramatic events in Yambuku and the slightly smaller crisis in southwestern Sudan, which was nonetheless large enough to account for 151 deaths. The Sudanese outbreak centered at a town near the Zairian border, five hundred miles northeast of Yambuku. It began among employees of a cotton factory, in the rafters of which roosted bats and on the floor of which skittered rats. The lethality was lower than in Zaire, "only" 53 percent, and laboratory analysis revealed that the Sudanese virus was genetically distinct enough from the virus in Zaire to be classified in a separate species. That species later became known, in careful taxonomic parlance, as *Sudan ebolavirus.* The official common name is simply Sudan virus, which lacks the frisson of the word "Ebola" but nonetheless denotes a dangerous, blazing killer. The version Karl Johnson and his colleagues had found at Yambuku, originally and still called Ebola virus, belongs to the species *Zaire ebolavirus.* This may seem confusing, but the accurate, up-to-date labels are important for keeping things straight. Eventually there would be five recognized species.

In 1977 a young girl died of hemorrhagic fever at a mission hospital in a village called Tandala, in northwestern Zaire. A blood sample taken after her death and sent unrefrigerated to the CDC yielded Ebola virus, not in cell cultures but only after inoculating live guinea pigs and then finding the virus replicating in their organs. (These were early days still in the modern field campaign against emerging viruses, and methodology was being extemporized to compensate for difficulties, such as keeping live virus frozen under rough field conditions in the tropics.) Karl Johnson again was part of the laboratory team; this seemed a logical extension of his work on the first outbreak, just a year earlier and two hundred miles east. But the nine-year-old girl, dead in Tandala, was an isolated case. Her family and friends remained uninfected. There was not even a hypothesis as to how she got sick. The later published report, with Johnson again as coauthor, only noted suggestively, in describing the girl's native area: "Contact with nature is intimate, with villages located in clearings of the dense rain forest or along the rivers of the savannah." Had she touched a dead chimpanzee, breathed rodent urine in a dusty shed, or pressed her lips to the wrong forest flower?

Two years later Sudan virus also resurfaced, infecting a worker at the same cotton factory where it had originally emerged. The worker was hospitalized, upon which he infected another patient, and by the time the virus finished ricocheting through that hospital, twenty-two people were dead. The case fatality rate was again high (65 percent), though lower than for Ebola virus. Sudan virus seemed to be not quite so lethal.

Then another decade passed before filoviruses made their next appearance, in another shape, in an unexpected place: Reston, Virginia.

You know about this if you've read *The Hot Zone*, Richard Preston's account of a 1989 outbreak of an Ebola-like virus among captive Asian monkeys at a lab-animal quarantine facility in suburban Reston, just across the Potomac from Washington, DC. Filovirus experts express mixed opinions about Preston's book, but there's no question that it did more than any journal article or newspaper story to make ebolaviruses infamous and terrifying to the general public. It also led to "a shower of funding," one expert told me, for virologists "who before didn't see a dime for their work on these exotic agents!" If this virus could massacre primates in their cages within a nondescript building in a Virginia office park, couldn't it go anywhere and kill anyone?

The facility in question was known as the Reston Primate Quarantine Unit and owned by a company called Hazelton Research Products, which was a division of Corning. The unfortunate monkeys were long-tailed macaques (*Macaca fascicularis*), an animal much used in medical research. They had arrived in an air shipment from the Philippines. Evidently they brought their filovirus with them, a lethal stowaway, like smallpox virus making its way through the crew of a sailing ship. Two macaques were dead on arrival, which wasn't unusual after such a stressful journey; but over the following weeks, within the building, many more died, which *was* unusual. Eventually the situation triggered alarm and the infective agent was recognized as an ebolavirus—*some* sort of ebolavirus, as yet unspecified. A team from USAMRIID came in, like a SWAT team in hazmat suits, to kill all the remaining macaques. Then they sealed the Reston Primate Quarantine Unit and sterilized it with formaldehyde gas. You can read Preston for the chilling details. There was great anxiety among the experts because this ebolavirus seemed to be traveling from monkey to monkey in airborne droplets; a leak from the building might therefore send it wafting out into Washington-area traffic. Was it lethal to humans as well as to macaques? Several staff members of the Quarantine Unit eventually tested positive for antibodies but—sigh of relief— those people showed no symptoms. Laboratory work revealed that the virus was similar to Ebola virus yet, like Sudan virus, different enough to be classified in a new species. It came to be known as Reston virus.

Notwithstanding that name, Reston virus seems to be native to the Philippines, not to suburban Virginia. Subsequent investigation of monkey-export houses near Manila, on the island of Luzon, found a sizable die-off of animals, most of which were infected with Reston virus, plus twelve people with antibodies to the virus. But none of the dozen Filipinos got sick. So the good news about Reston virus, derived both from the 1989 US scare and from retrospective research on Luzon, is that it doesn't seem to cause illness in humans, only in monkeys. The bad news is that no one understands why.

Apart from Reston virus, ebolaviruses in the wild remain an African phenomenon. But the next emergence, in November 1992, added yet another point to the African map. Chimpanzees began dying at a forest refuge in Côte d'Ivoire, West Africa. The refuge, Taï National Park, lying near Côte d'Ivoire's border with Liberia, encompassed one of the last remaining areas of primary rainforest in that part of Africa. It harbored a rich diversity of animals, including several thousand chimpanzees.

One community of those chimps had been followed and studied for thirteen years by a Swiss biologist named Christophe Boesch. During the 1992 episode, Boesch and his colleagues noticed a sudden drop in the population—some chimps died, others disappeared—but the scientists didn't detect a cause. Then, in late 1994, eight more carcasses turned up over a short period of time, and again other animals went missing. Two of the chimp bodies, only moderately decayed, were cut open and examined by researchers at Taï. One of them proved to be teeming with an Ebola-like agent, though that wasn't apparent at the time. During the necropsy, a thirty-four-year-old female Swiss graduate student, wearing gloves but no gown, no mask, became infected. Infected how? There wasn't any obvious moment of fateful exposure, no slip of the scalpel, no needlestick mishap. Probably she got chimp blood onto a broken patch of skin—a small scratch?—or caught a gentle splash of droplets in the face. Eight days later, the woman started shivering.

She took a dose of malaria medicine. That didn't help. She was moved to a clinic in Abidjan, Côte d'Ivoire's capital, and there treated again for malaria. Her fever continued. On day 5 came vomiting and diarrhea, plus a rash that spread over her whole body. On day 7 she was carried aboard an ambulance jet and flown to Switzerland. Now she *was* wearing a mask, and so were the doctor and the nurse in attendance. But no one knew what ailed her. Dengue fever, hantavirus infection, and typhoid were being considered, and malaria still hadn't been ruled out. (Ebola wasn't at the top of the list because it had never been seen in Côte d'Ivoire.) In Switzerland, hospitalized within a double-door isolation room with negative air pressure, she was tested for a whole menu of nasty things, including Lassa fever, Crimean-Congo hemorrhagic fever, chikungunya, yellow fever, Marburg virus disease, and now, yes, Ebola virus disease. The last of those possibilities was investigated using three kinds of assays, each one specific: for Ebola virus, for Sudan virus, for Reston virus. No positive results. The antibodies in those assays didn't recognize the virus, whatever it was, in her blood.

The laboratory sleuths persisted, designing a fourth assay that was more generalized—comprehensive for the whole group of ebolaviruses. Applied to her serum, that one glowed, a positive, announcing the presence of antibodies to an ebolavirus of some sort. So the Swiss woman was the world's first identified victim of what became known as Taï Forest virus. The chimpanzee she had necropsied, its tissues tested later, was the second victim, recognized posthumously.

Unlike the chimp, she survived. After another week, she left the hospital. She had lost thirteen pounds and her hair later fell out, but otherwise she was okay. Besides being the initial case of Taï Forest virus infection, the Swiss woman holds one other distinction: She is the first person known to have carried an ebolavirus infection off the African continent. Events in 2014 have shown that she wasn't the last.

6

Ebolavirus spillovers continued throughout the 1990s and into the twenty-first century, sporadic and scattered enough to make field research difficult, frequent enough to keep some scientists focused and some public health officials worried. In 1995, soon after the Côte d'Ivoire episode, it was Ebola virus in Kikwit, about which you've read. Six months after that outbreak, as you'll also recall, the new one began at Mayibout 2. What I haven't yet mentioned about Mayibout 2 is that, though the village lies in Gabon, the virus was Ebola as known originally from Zaire, which seems to be the most broadly distributed of the group. And again at the timber camp near Booué, Gabon, it was Ebola virus.

Also that year, 1996, Reston virus reentered the United States by way of another shipment of Philippine macaques. Sent from the same export house near Manila that had shipped the original sick monkeys to Reston, Virginia, these went to a commercial quarantine facility in Alice, Texas, near Corpus Christi. One animal died and, after it tested positive for Reston virus, forty-nine others housed in the same room were "euthanized" as a precaution. (Most of those, tested posthumously, were negative.) Ten employees who had helped unload and handle the monkeys were also screened for infection, and they also tested negative, but none of them were euthanized.

Uganda became the next known locus of the virus in Africa, with an outbreak of Sudan virus that began near the northern town of Gulu in August 2000. Northern Uganda shares a border with what in those days was southern Sudan, and it wasn't surprising that Sudan virus might cross or straddle that border. Cross it how, straddle it how? By way of the individual movements or the collective distribution of the reservoir host, identity unknown. This is a pointed example of why solving the reservoir mystery is important: If you know which animal harbors a certain virus and where that animal lives—and conversely, where it *doesn't* live—you know where the virus may next spill over, and where it probably won't. That provides some basis for focusing your vigilance. If the reservoir is a rodent that lives in the forests of southwestern Sudan but not in the deserts of Niger, the goat herders of Niger can relax. They have other things to worry about.

In Uganda, unfortunately, the 2000 spillover led to an epidemic of Sudan virus infections that spread from village to village, from hospital to hospital, from the north of the country to the southwest, killing 224 people.

The case fatality rate was again "only" 53 percent, exactly what it had been in the first Sudanese outbreak, back in 1976. This precise coincidence seemed to reflect a significant difference in virulence between Sudan and Ebola viruses. Their difference, in turn, might reflect different evolutionary adjustments to humans as a secondary host (though random happenstance is also a possible explanation). Many factors contribute to the case fatality rate during an outbreak, including diet, economic conditions, public health in general, and the medical care available in the location where an outbreak occurs. It's hard to isolate the inherent ferocity of a virus from those contextual factors. What can be said, though, is that Ebola virus *appears* to be the meanest of the four ebolaviruses you've heard about, as gauged by its effect on human populations. Taï Forest virus can't reliably be placed on that spectrum at all, not yet—for lack of evidence. Having infected just one known human (or possibly two, counting an unconfirmed later case) and killed none, Taï Forest virus may be less prone to spillover. It may or may not be less lethal; one case, like one roll of the dice, proves nothing about what's likely to emerge as numbers grow larger. Then again, Taï Forest virus might also be spilling more frequently but inconsequentially—infecting people yet not causing notable illness. No one has screened the populace of Côte d'Ivoire to exclude that possibility.

The role of evolution in making Taï Forest virus (or any virus) less virulent in humans is a complicated matter, not easily deduced from simple comparison of case fatality rates. Sheer lethality may be irrelevant to the virus's reproductive success and long-term survival, the measures by which evolution keeps score. Remember, the human body isn't the primary habitat of ebolaviruses. The reservoir host is.

Like other zoonotic viruses, ebolaviruses have probably adapted to living tranquilly within their reservoir (or reservoirs), replicating steadily but not abundantly and causing little or no trouble. Spilling over into humans, they encounter a new environment, a new set of circumstances, often causing fatal devastation. And one human can infect another, through direct contact with bodily fluids. But the chain of ebolavirus infection, at least so far, has never continued through many thousands of successive cases, great distances, or long stretches of time. Some scientists use the term "dead-end host" to describe humanity's role in the lives and adventures of ebolaviruses. What the term implies is this: Outbreaks have been contained and terminated; in each situation the virus has come to a dead end, leaving no offspring. Not the virus in toto throughout its range, of course, but that *lineage* of virus, the one that has spilled over, betting everything on this gambit—it's gone, kaput. It's an evolutionary loser. It hasn't caught hold to become an endemic disease within human populations. It hasn't caused a huge epidemic. Ebolaviruses, judged by experience so far, fit that pattern. Careful medical procedures (such as barrier nursing by way of isolation wards, examination gloves, gowns, masks, and disposable needles and syringes) usually stop them. Sometimes simpler methods can bring a local spillover to a dead end too. This has probably happened more times than we'll ever know. Advisory: If your husband catches an ebolavirus, give him food and water and love and maybe prayers but keep your distance, wait patiently, hope for the best—and, if he dies, don't clean out his bowels by hand. Better to step back, blow a kiss, and burn the hut.

This business about dead-end hosts is the conventional wisdom. It applies to the ordinary course of events. But there's another perspective to consider. Zoonoses by definition involve events beyond the ordinary, and the scope of their consequences can be extraordinary too. Every spillover is like a sweepstakes ticket, bought by the pathogen, for the prize of a new and more grandiose existence. It's a long-shot chance to transcend the dead end. To go where it hasn't gone and be what it hasn't been. Sometimes the bettor wins big.

7

In late 2007 a fifth ebolavirus emerged, this one in western Uganda.

On November 5, 2007, the Ugandan Ministry of Health received a report of twenty mysterious deaths in Bundibugyo, a remote district along the mountainous border with the Democratic Republic of the Congo (the new name, as of 1997, for what had been Zaire). An acute infection of some unknown sort had killed those twenty people, abruptly, and put others at risk. Was it a rickettsial bacterium, such as the one that causes typhus? An ebolavirus was another possibility, but considered less likely at first, because few of the patients hemorrhaged. Blood samples were gathered quickly, flown to the CDC in Atlanta, and tested there, using both a generalized assay that might detect any form of ebolavirus and specific assays for each of the known four. Although the specific tests were all negative, the general test rang up some positives. So on November 28, the CDC informed Ugandan officials: It's an ebolavirus, all right, but not one we've ever seen.

Further laboratory work established that this new virus was at least 32 percent different genetically from any of the other four. It became Bundibugyo virus. Soon a CDC field team arrived in Uganda to help respond to the outbreak. As usual in such situations, their efforts along with those of the national health authorities involved three tasks: caring for patients, trying to prevent further spread, and investigating the nature of the disease. The eventual tally was 116 people infected, of whom 39 died.

Also as usual, the scientific team later published a journal article, in this case announcing the discovery of a new ebolavirus. First author on the paper was Jonathan S. Towner, a molecular virologist at the CDC with field experience in the search for reservoirs. Besides guiding the lab work, he went to Uganda and did a stint with the response team. The Towner paper contained a very interesting statement, as an aside, concerning the five ebolaviruses: "Viruses of each species have genomes that are at least 30–40% divergent from one another, a level of diversity that presumably reflects differences in the ecologic niche they occupy and in their evolutionary history." Towner and company suggested that some of the crucial differences between one ebolavirus and another—including the differences in lethality—might be related to where and how they live, where and how they *have* lived, within their reservoir hosts.

The events in Bundibugyo left many Ugandans uneasy. And they were entitled to their uneasiness: Uganda now held a sorry distinction as the only country on Earth that had suffered outbreaks of two different ebolaviruses (Sudan virus at Gulu in 2000, Bundibugyo virus in 2007), as well as outbreaks of both Ebola virus disease and Marburg virus disease, caused by another filovirus, within a single year. (The creepy circumstances of the Marburg spillover, at a gold mine called Kitaka in June 2007, are part of a story I'll come to in its turn.) Given such national ill fortune, it's not surprising that there were rumors, stories, and anxieties circulating among Ugandans, in late 2007, that made tracing genuine ebolavirus leads all the more difficult.

A pregnant woman, showing signs of hemorrhagic fever,

delivered her baby and then died. The baby, left in the care of a grandmother, soon died too. That was sad but not peculiar; orphaned infants often die in the hard conditions of a village. More notable was that the grandmother also died. An ape (unidentified, but chimp or gorilla) reportedly bit a domestic goat, infecting it; the goat was slaughtered in due course, skinned by a thirteen-year-old boy, and then the boy's family began falling ill. No, a dead monkey was eaten. No, bats were eaten. Mostly these tales couldn't be substantiated, but their currency and their general themes reflected a widespread, intuitive comprehension of zoonoses: Relations between humans and other animals, wild or domestic, must somehow lie at the root of the disease troubles. In early December, and then again in January 2008, came reports of suspicious animal deaths (monkeys and pigs) in outlying regions of the country. One of those reports also involved dogs that died after being bitten by the sickened monkeys. Was it an epidemic of rabies? Was it Ebola? The Ministry of Health sent people to collect specimens and investigate.

"Then there was a new epidemic—of fear," said Dr. Sam Okware, Commissioner of Health Services, when I visited him in Kampala a month later. Among Dr. Okware's other duties, he served as chairman of the national Ebola virus task force. "That was the most difficult to contain," he said. "There was a new epidemic—of panic."

These are remote places, he explained. Villages, settlements, small towns surrounded by forest. The people feed themselves mostly on wildlife. During the Bundibugyo outbreak, residents of that area were shunned. Their economy froze. Outsiders wouldn't accept their money, scared that it carried infection. Population drained from the major town. The bank closed. When patients recovered (if they were lucky enough to recover) and went home from the hospital, "again they were shunned. Their houses were burned." Dr. Okware was a thin, middle-aged man with a trim mustache and long, gesticulant hands that moved through the air as he spoke of Uganda's traumatic year. The Bundibugyo outbreak, he said, was "insidious" more than dramatic, smoldering ambiguously while health officials struggled to comprehend it. There were still five questions pending, he said, and he began to list them: (1) Why were only half of the members of each household affected? (2) Why were so few hospital workers affected, compared to other Ebola outbreaks? (3) Why did the disease strike so spottily within the Bundibugyo district, hitting some villages but not others? (4) Was the infection transmitted by sexual contact? After those four he paused, momentarily unable to recall his fifth pending question.

"The reservoir?" I suggested. Yes, that's it, he said: *What's the reservoir?*

Bundibugyo virus in Uganda, 2007, completes the outline sketch of ebolavirus classification as presently known. Four different ebolaviruses are scattered variously across sub-Saharan Africa and had emerged, as of that year, from their reservoir hosts to cause human disease (as well as gorilla and chimpanzee deaths) in six different countries: Sudan, Gabon, Uganda, Côte d'Ivoire, the Republic of the Congo, and the Democratic Republic of the Congo. (This was the state of distributional knowledge until the West Africa outbreak of 2014, which I'll discuss in the epilogue, below.) A fifth ebolavirus seems to be endemic to the Philippines, and to have traveled from there several times to the United States in infected macaques. But how did it get to the Philippines, if the ancestral origin of ebolaviruses is equatorial Africa? Could it have arrived there in one soaring leap, leaving no traces in between? From southwestern Sudan to Manila is almost seven thousand miles as the bat flies. But no bat can fly that far without roosting. Are ebolaviruses more broadly distributed than we suspect? Should scientists start looking for them in India, Thailand, and Vietnam? (Antibodies for Ebola virus or something closely related seem to have been found among people in Madagascar. Go figure.) Or did Reston virus get to the Philippines the same way Taï Forest virus got to Switzerland and Ebola virus to Johannesburg—by airplane?

If you contemplate all this from the perspective of biogeography (the study of which creatures live where on planet Earth) and phylogeny (the study of evolving lineages), one thing becomes evident: The current scientific understanding of ebolaviruses constitutes pinpricks of light against a dark background.

8

People in the villages where Ebola struck—the survivors, the bereaved, the scared but lucky ones not directly affected— had their own ways of understanding this phenomenon, and one way was in terms of malevolent spirits. In a single word, which loosely encompasses the variety of beliefs and practices seen among different ethnic and language groups and is often used to explain rapid death of adults: sorcery.

The village of Mékouka, on the upper Ivindo River in northeastern Gabon, offers an instance. Mékouka was one of the gold camps in which the outbreak of 1994 got its start. Three years later, a medical anthropologist named Barry Hewlett, an American, visited there to learn from the villagers themselves how they had thought about and responded to the outbreak. Many local people told him, using a term from their Bakola language, that this Ebola thing was *ezanga,* meaning some sort of vampirism or evil spirit. Asked to elaborate, one villager explained that ezanga are "bad human-like spirits that cause illness in people" as retribution for accumulating material goods and not sharing. (This wouldn't seem to apply to that man on the upper Ivindo, in 1994, who shared his tainted gorilla meat before he died.) Ezanga could even be summoned and targeted at a victim, like casting a hex. Neighbors or acquaintances, envious of the wealth or power someone has amassed, could send ezanga to gnaw at the person's internal organs, making him sick unto death. That's why gold miners and timber-company employees suffered such high risk of Ebola, Hewlett was told. They were envied and they didn't share.

Barry Hewlett had investigated the Mékouka outbreak in retrospect, months after the events occurred. Still fascinated by the subject, and concerned that an important dimension was being omitted by the more clinical methods of research and response, he got himself to the scene in Gulu, Uganda, in late 2000, while that outbreak was still going on. He found that the predominant ethnic group there, the Acholi, were also inclined to attribute Ebola virus disease to supernatural forces. They believed in a form of malign spirit, called *gemo*, that sometimes swept in like the wind to cause waves of sickness and death. Ebola wasn't their first gemo. The Acholi previously suffered epidemics of measles and smallpox, Hewlett learned, and those were likewise explained. Several elders told Hewlett that disrespect for the spirits of nature could bring on a gemo.

Once a true gemo was recognized, as distinct from a lesser spate of illness in the community, Acholi cultural knowledge dictated a program of special behaviors, some of which were quite appropriate for controlling infectious disease, whether you believed it was caused by spirits or by a virus. These behaviors included quarantining each patient in a house apart from other houses; relying on a survivor of the epidemic (if there were any) to provide care to each patient; limiting movement of people between the affected village and others; abstaining from sexual relations; not eating rotten or smoked meat; and suspending the ordinary burial practices, which would involve an open casket and a final "love touch" of the deceased by each mourner, filing up for that purpose. Dancing was also prohibited. Such traditional Acholi strictures (along with intervention by the Uganda Ministry of Health and support from the CDC, Médecins sans Frontières, and the WHO) may have helped suppress the Gulu outbreak.

"We have a lot to learn from these people," Barry Hewlett told me, one day in Gabon, "as to how they've responded to these epidemics over time." Modern society has lost that sort of ancient, painfully acquired accumulation of cultural knowledge, he said. Instead we depend on the disease scientists. Molecular biology and epidemiology are useful, but other traditions of knowledge are useful too. "Let's listen to what people are saying here. Let's find out what's going on. They've been living with epidemics for a long time."

Hewlett is a gentle-spirited man with a professorship at Washington State University and two decades of field experience in Central Africa. By the time I met him, at an international ebolavirus conference in Libreville, we had each visited one other village famed for suffering the disease—a place called Mbomo, in the Republic of the Congo, along the western edge of Odzala National Park. Mbomo lies not far from the Mambili River and the Moba Bai complex, where I had watched Billy Karesh trying to dart gorillas. The outbreak around Mbomo began in December 2002, probably among hunters who handled infected gorillas or duikers, and spread throughout an area that encompassed at least two other villages. A large difference between Hewlett's experience in Mbomo and mine was that he arrived during the outbreak. The grease was still flaming in the pan when he made his inquiries.

One early patient, Hewlett learned, was pulled out of the village clinic because his family disbelieved the Ebola diagnosis and preferred relying on a traditional healer. After that patient died at home, unattended by medical personnel and uncured by the healer, things got testy. The healer pronounced that this man had been poisoned by sorcery and that the perpetrator was his older brother, a successful man working in a nearby village. The older brother was a teacher who had "risen" to become a school inspector and didn't share the good fortune with his family. So again, as with ezanga among the Bakola people in northeastern Gabon, there were jealous animosities underlying the accusations of sorcery. Then another brother died, and a nephew, at which point family members burned the older brother's Mbomo house and sent a posse to kill him. They were stopped by the police. The older brother, though now taken for an evil magus, escaped vengeance. Then community relations deteriorated generally as more victims died from the invisible terror, with no cure available, no satisfactory explanation, to a point where anyone who looked out of the ordinary or above the crowd became suspect.

Another element of the dangerous brew in and around Mbomo was a mystic secret society, *La Rose Croix*, more familiar (if barely) to you and me as Rosicrucianism. It's an international organization that has existed for centuries, mostly devoted to esoteric study, but in this part of the Congo it had a bad reputation, akin to sorcery. Four teachers within one nearby village were members, or were thought to be members—and these teachers had been telling children about Ebola virus before the outbreak occurred. That led some traditional healers to suspect that the teachers had advance knowledge—supernatural knowledge—of the outbreak. Something had to be done, yes? On the day before Barry Hewlett and his wife arrived in Mbomo, the four teachers were murdered with machetes while they worked in their crop fields.

Soon afterward, the disease outbreak expanded to include so many community members that sorcery no longer seemed a plausible explanation to local people. The alternative was *opepe*, an epidemic, Mbomo's equivalent (in Kota, one of the local languages) to what Barry Hewlett had heard about, from the Acholi, as gemo. "This illness is killing everyone," one local man told the Hewletts, and therefore it couldn't be sorcery, which targets individual victims or their families. By early June 2003, there had been 143 cases in Mbomo and the surrounding area, with 128 deaths. That's a case fatality rate of 90 percent, at the top of the range even for Ebola virus.

With their deep interest in local explanations and their patient listening methods, the Hewletts heard things that wouldn't fit within the multiple-choice categories of an epidemiological questionnaire. Another of their informants, an Mbomo woman, declared: "Sorcery does not kill without reason, does not kill everybody, and does not kill gorillas or other animals." Oh, yes, again gorillas. That was another aspect of the Mbomo brew—everyone knew there were dead apes in the forest all roundabout. They had died at the Lossi sanctuary. They had died, so far as Billy Karesh could tell, at Moba Bai. Carcasses had been seen in the environs of Mbomo itself. And, as the woman said, sorcery didn't apply to gorillas.

9

When a silverback gorilla dies of Ebola, he does it beyond the eyes of science and medicine. No one is there in the forest to observe the course of his agony, with the possible exception of other gorillas. No one takes his temperature or peers down his throat. When a female gorilla succumbs to Ebola, no one measures the rate of her breathing or checks for a telltale rash. Thousands of gorillas may have been killed by the virus but no human has ever attended one of those deaths—not even Billy Karesh, not even Alain Ondzie. A small number of carcasses have been found, some of which have tested positive for Ebola antibodies. A larger number of carcasses have been seen and reported by casual witnesses, in Ebola territory at Ebola times, but because the forest is a hungry place, most of those carcasses could never be inspected and sampled by scientific researchers. The rest of what we know about Ebola's effect on gorillas is inferential: Many of them—major portions of some regional populations, such as the ones at Lossi, Odzala, and Minkébé—have disappeared. But nobody knows just how Ebola virus affects the gorilla body.

With humans it's different. The numbers I've mentioned above offer one gauge of that difference: 245 fatal cases during the outbreak at Kikwit, another 224 at Gulu, 128 in and around Mbomo, et cetera. The total of known human fatalities from ebolavirus infection, from the discovery of the first ebolaviruses in 1976 through the end of 2012, was about 1,580. The West African outbreak of 2014 (which seems to have begun in southern Guinea, as early as December 2013) has more than doubled the total as of this writing, with no end yet in sight; the death count is rising so quickly, right now, that it's pointless for me to print a number. Although the suffering has been awful, the total is still relatively low compared to the tolls taken by such widespread and relentless global afflictions as malaria and tuberculosis, or to the great waves of death brought by the various influenzas.

But it's high enough to have yielded a significant body of data. Furthermore, doctors and nurses have seen many of those victims die. So the medical profession knows a good bit about the range of symptoms and the pathological effects produced on a human body during death by ebolavirus infection. It's not quite like you might think.

If you devoured *The Hot Zone* when it was published, as I did, or if you have been secondarily exposed to its far-reaching influence on public impressions about ebolaviruses, you may carry some wildly gruesome notions. Richard Preston is a vivid writer, a skillful writer, an industrious researcher, and it was his purpose to make a truly horrible disease seem almost preternaturally horrific. You may recall his depiction of a Sudanese hospital in which the virus "jumped from bed to bed, killing patients left and right," creating dementia and chaos, and not only killing patients but causing them to bleed profusely as they died, liquefying their organs, until "people were dissolving in their beds." You may have shuddered at Preston's statement that Ebola virus in particular "transforms virtually every part of the body into a digested slime of virus particles." You may have paused before turning the page when he told you that, after death, an Ebola-infected cadaver "suddenly deteriorates," its internal organs deliquescing in "a sort of shock-related meltdown." You may not have noticed that meltdown was a metaphor, meaning dysfunction, not actual melting. Or maybe it wasn't. At a later point, bringing another filovirus into the story, Preston mentioned a French expatriate, living in Africa, who "essentially melts down with Marburg virus while traveling on an airplane." You may remember one phrase in particular, as Preston described victims in a darkened Sudanese hut: comatose, motionless, and "bleeding out." That seemed to be so different from just "bleeding." It suggested a human body draining away in a gush. There was also the statement that Ebola causes a victim's eyeballs to fill up with blood, bringing blindness and more. "Droplets of blood stand out on the eyelids: You may weep blood. The blood runs from your eyes down your cheeks and refuses to coagulate." The mask of red death—where medical reporting meets Edgar Allan Poe.

It's my duty to advise that you need not take these descriptions quite literally—at least, not as the typical course of a fatal case of Ebola virus disease. Expert testimony, some published and some spoken, tempers Preston on several of these more lurid points, without diminishing the terribleness of Ebola in terms of real suffering and death. Pierre Rollin, for instance, now deputy chief of the Viral Special Pathogens Branch of the CDC, is one of the world's most experienced ebolavirus hands. He worked at the Pasteur Institute in Paris before moving to Atlanta, and has been a member of response teams to many Ebola and Marburg outbreaks over the past fifteen years, including those at Kikwit and Gulu. When I asked him, during an interview in his office, about the public perception that this disease is extraordinarily bloody, Rollin interrupted me genially to say: "—which is bullshit." When I mentioned the descriptions in Preston's book, Rollin mockingly said, "They melt, splash on the wall," and gave a frustrated shrug. Mr. Preston could write what he pleased, Rollin added, so long as the product was labeled fiction. "But if you say it's a true story, you have to speak to the true story, and he didn't. Because it was much more exciting to have blood everywhere and scaring everywhere." A few patients do bleed to death, Rollin said, but "they don't explode, and they don't melt." In fact, he said, the conventional term then in use, "Ebola hemorrhagic fever," was itself a misnomer because more than half the patients don't bleed at all. They die of other causes, such as respiratory distress and shutdown (but not dissolution) of internal organs. It's for just these reasons, as cited by Rollin, that the WHO has switched its own terminology from "Ebola hemorrhagic fever" to "Ebola virus disease."

Karl Johnson, one of the pioneers of Ebola outbreak response, whose credentials I've already sketched, offered a similar but even more pointed reaction, expressed with his usual candor. We were talking—in my own office, this time—during one of his periodic trips to Montana for fly-fishing. We had become friends and he had coached me a bit, informally, on how to think about zoonotic viruses. Finally I got him to sit for an interview, and *The Hot Zone* inescapably came up. Waxing serious, Karl said: "Bloody tears is bullshit. Nobody has ever had bloody tears." Furthermore, Karl noted, "People who die are not formless bags of slime." Johnson also concurred with Pierre Rollin that the bloodiness angle has been oversold. If you want a really bloody disease, he said, look at Crimean-Congo hemorrhagic fever. Ebola is bad and lethal, sure, but not bad and lethal precisely that way.

In the real world, as described in the scientific literature, the list of major symptoms of Ebola virus disease goes like this: abdominal pain, fever, headache, sore throat, nausea and vomiting, loss of appetite, arthralgia (joint pain), myalgia (muscle pain), asthenia (weakness), tachypnea (rapid breathing), conjunctival injection, and diarrhea. Conjunctival injection means pink eye, not bloody tears. All these symptoms tend to show up in many or most fatal cases. Additional symptoms including chest pain, hematemesis (vomiting of blood), bleeding from the gums, bloody stools, bleeding from needle-puncture sites, anuria (inability to pee), rash, hiccups, and ringing in the ears have appeared in a smaller fraction of cases. During the Kikwit outbreak, 59 percent of all patients didn't bleed noticeably at all, and bleeding in general was no indicator of who would or wouldn't survive. Rapid breathing, urine retention, and hiccups, on the other hand, were ominous signals that death would probably come soon. Among those patients who did bleed, blood loss never seemed massive, except among pregnant women who spontaneously aborted their fetuses. Most of the nonsurvivors died stuporous and in shock. Which is to say: Ebola virus generally killed with a whimper, not with a bang or a splash.

Despite all these data, gathered amid woeful and dangerous conditions while the primary mission was not science but saving lives, even the experts aren't sure exactly *how* the virus typically causes death. "We don't know the mechanism," Pierre Rollin told me. He could point to liver failure, to kidney failure, to breathing difficulties, to diarrhea, and in the end it often seemed that multiple causes were converging in an unstoppable cascade. Karl Johnson voiced similar uncertainty, but mentioned that the virus "really goes after the immune system," shutting down production of interferon, a class of proteins essential to immune response, so that "nothing stops the continued replication of the virus."

This idea of immune suppression by ebolaviruses has also appeared lately in the literature, along with speculation that it might allow catastrophic overgrowth of a patient's natural populations of bacteria, normally resident in the gut and elsewhere, as well as unhindered replication of the virus itself. Runaway bacterial growth might in turn put blood into the urine and feces, and even lead to "intestinal destruction," according to one source. Maybe that's what Preston had in mind when he wrote about liquefied organs and people dissolving in their beds. If so, he was blurring the distinction between what Ebola virus does and what garden-variety bacteria can do in the absence of a healthy immune system keeping them cropped. But, hey, don't we all like a dramatic story better than a complicated one?

Still another aspect of the pathology of Ebola virus disease is a phenomenon called disseminated intravascular coagulation, familiar to the medical community as DIC. It's also known as consumptive coagulopathy (if that helps you), because it involves consumption of too much of the blood's coagulating capacity in a misdirected way. Billy Karesh had told me about DIC as we boated down the Mambili River after our gorilla stakeout. Disseminated intravascular coagulation, he explained, is a form of pathological blood sludge, in which the normal clotting factors (coagulation proteins and platelets) are pulled out to form tiny clots along the insides of blood vessels throughout the victim's body, leaving little or no coagulation capacity to prevent leakage elsewhere. As a result, blood may seep from capillaries into a person's skin, forming bruiselike purple marks (hematomas); it may dribble from a needle puncture that seems never to heal, or it may leak into the gastrointestinal tract or the urine. Still worse, the mass aggregation of small clots in the vessels may block blood flow to the kidneys or the liver, causing organ failure as often seen with Ebola.

At least that was the understanding of DIC's role in Ebola virus disease at the time Karesh alerted me to it. More recently, Karl Johnson and others have begun questioning whether the immune-shutdown effect that the virus somehow achieves, and the consequent blossoms of bacteria, might better explain some of the damage formerly blamed on DIC. "When it was first discovered, DIC, da da da, was the key to everything in hemorrhagic fever," Johnson told me, again cheerily dismissive of conventional wisdom. Now, he said, he was reading a hell of a lot less about DIC in the literature.

Ebola virus is still an inscrutable bug in more ways than one, and Ebola virus disease is still a mystifying affliction as well as a ghastly, incurable one—with or without DIC, with or without melting organs and bloody tears. "I mean, it's awful," Johnson stressed. "It really, really is." He had seen it almost before anyone else, under especially mystifying conditions—in Zaire, 1976, before the virus even had a name. But the thing hasn't changed, he said. "And frankly, everybody in the world is much too afraid of it, including the medical fraternity worldwide, to really want to try and study it." To study its effect on a living, struggling human body, he meant. To do that, you would need the right combination of hospital facilities, BSL-4 facilities, dedicated and expert professionals, and circumstances. You couldn't do it during the next outbreak at a mission clinic in an African village. You would need to bring Ebola virus into captivity—into a research situation, under highly controlled scrutiny—and not just in the form of frozen samples. You would need to study a raging infection inside somebody's body.

That isn't easy to arrange. He added: "We haven't had an Ebola patient yet in the United States." He was speaking in 2008. But for everything that happens, there is a first time. For the United States, it occurred in 2014, when two infected Americans were brought back from West Africa to an isolation suite at Emory University Hospital. And there were precursors.

10

England had its first case of Ebola virus disease in 1976. Russia had its first case (that we know of) in 1996. Unlike the Swiss woman who did the chimp necropsy in Côte d'Ivoire, these two unfortunate people didn't pick up their infections during African fieldwork and come home prostrate in an ambulance jet. Their exposure derived from laboratory accidents. Each of them suffered a small, fateful, self-inflicted injury while doing research.

The English accident occurred at Britain's Microbiological Research Establishment, a discreetly expert institution within a high-security government compound known as Porton Down, not far from Stonehenge in the rolling green countryside southwest of London. Think of Los Alamos, but tucked into the boonies of pastoral England instead of the mountains of New Mexico, and with bacteria and viruses in place of uranium and plutonium as the strategic materials of interest. In its early years, beginning in 1916, Porton Down was an experiment station for the development of chemical weapons such as mustard gas; during World War II, its scientists worked also on biological weapons derived from anthrax and botulin bacteria. But eventually, at Porton Down as at USAMRIID, with changing political circumstances and government scruples, the emphasis shifted to defense—that is, research on countermeasures against biological and chemical weapons. That work involved high-containment facilities and techniques for studying dangerous new viruses, and therefore qualified Porton Down to offer assistance in 1976, when the WHO assembled a field team to investigate a mysterious disease outbreak in southwestern Sudan. Deep-frozen blood samples from desperately ill Sudanese patients arrived for analysis—at about the same time, during that fretful autumn, as blood samples from Yambuku went to the CDC. The field people were asking the laboratory people to help answer a question: What *is* this thing? It hadn't yet been given a name.

One of the lab people at Porton Down was Geoffrey S. Platt. On November 5, 1976, in the course of an experiment, Platt filled a syringe with homogenized liver from a guinea pig that had been infected with the Sudanese virus. Presumably he intended to inject that fluid into another test animal. Something went amiss, and instead he jabbed himself in the thumb.

Platt didn't know exactly what pathogen he had just exposed himself to, but he knew it wasn't good. The fatality rate from this unidentified virus, as he must have been aware, was upwards of 50 percent. Immediately he peeled off his glove, plunged his thumb into a hypochlorite solution (bleach-like stuff, which kills virus) and tried to squeeze out a drop or two of blood. None came. He couldn't even see a puncture. That was a good sign if it meant there *was* no puncture, a bad sign if it meant a little hole sealed tight. The tininess of Platt's wound, in light of subsequent events, testifies that even a minuscule dose of an ebolavirus is enough to cause infection, at least if that dose gets directly into a person's bloodstream. Not every pathogen is so potent. Some require a more sizable foothold. Ebolaviruses have force but not reach. You can't catch one by breathing shared air, but if a smidgen of the virus gets through a break in your skin (and there are always tiny breaks), God help you. In the terms used by the scientists: It's not very contagious but it's highly infectious. Six days after the needle prick, Geoffrey Platt got sick.

At the start he merely felt nauseous and exhausted, with abdominal pain. Given the circumstances, though, his malaise was taken very seriously. He was admitted to a special unit for infectious diseases at a hospital near London and, within that unit, put into a plastic-walled isolator tent under negative air pressure. The historical records don't mention it but you can be sure his nurses and doctors wore masks. He was given injections of interferon, to help stimulate his immune system, and blood serum (flown up from Africa) that had been drawn from a recovered Ebola patient to supply some borrowed antibodies. On the fourth day, Platt's temperature spiked and he vomited. This suggested the virus was thriving. For the next three days, his crisis period, he suffered more vomiting, plus diarrhea, and a spreading rash; his urine output was low; and a fungal growth in his throat hinted at immune failure. All these were gloomy signs. Meanwhile he was given more serum. Maybe it helped.

By the eighth day, Platt's vomiting and diarrhea had ended. Two days later, the rash began to fade and the fungus was under control. He had been lucky, perhaps genetically, as well as privileged to receive optimal medical care. The virus disappeared from his blood, from his urine, and from his feces (though it lingered awhile in his semen; apparently he promised doctors that he wouldn't make that a risk issue for anyone else). He was taken out of the isolator. Eventually he went home. He had lost weight, and during the long, slow convalescence much of his hair fell out. But like the Swiss woman, he survived.

The Russian researcher, in 1996, wasn't so lucky. Her name, as given in one Russian news account (but unspoken in the western medical literature), was Nadezhda Alekseevna Makovetskaya. Employed at a virological institute under the Ministry of Defense, she had been working on an experimental therapy against Ebola virus disease, derived from the blood serum of horses. Horses aren't highly susceptible to Ebola, which is why they are used to make antibodies. Testing the efficacy of this treatment required exposing additional horses. "It is difficult to describe working with a horse infected with Ebola," according to the dry, cautious statement from Russia's chief biowarfare man at the time, a lieutenant general named Valentin Yevstigneyev, in the Ministry of Defense. No doubt he was right about that. A horse can be nervous and jumpy, even if it's not suffering convulsions. Who would want to get close with a needle? "Under normal conditions this animal is difficult to manage and we had to work in special protective gear," said General Yevstigneyev. What he meant by "we" might be broadly interpreted. He was a high officer and military bureaucrat, not likely pulling the latex mitts onto his own hands. "One false step, one torn glove and the consequences would be grave." Makovetskaya had evidently taken that false step. Or maybe it wasn't her mistake so much as the twitch of a sensitive gelding. "She tore her protective gloves but concealed it from the leadership," by General Yevstigneyev's unsympathetic account, "since it happened just before the New Year holidays." Was he implying that she hadn't wanted to miss seasonal festivities while sitting in quarantine? He didn't mention a needlestick, or a scratch, or an open cut beneath the torn glove, though some such misfortune must have been involved. "As a result, by the time she turned to a doctor for help it was too late." The details of Makovetskaya's symptoms and death remain secret.

Another Russian woman stuck herself with Ebola in May 2004, and about this case slightly more is known. Antonina Presnyakova was a forty-six-year-old technician working at a high-security viral research center called Vektor (which sounds like something from Ian Fleming) in southwestern Siberia. Presnyakova's syringe carried blood from a guinea pig infected with Ebola virus. The needle went through two layers of gloves into her left palm. She immediately entered an isolation clinic, developed symptoms within a few days, and died at the end of two weeks.

These three cases reflect the inherent perils of doing laboratory research on such a lethal, infectious virus. They also suggest the context of concerns that surrounded America's closest approach to a home-grown case of Ebola. This one occurred also in 2004, just months before the death of Antonina Presnyakova.

11

Kelly L. Warfield grew up in a suburb of Frederick, Maryland, not many miles from Fort Detrick, the US Army base devoted to medical research and biodefense within which sits USAMRIID. She was a local girl, bright and curious, whose mother owned a convenience store just outside the Fort Detrick gate. Helping her mom since she was a middle-schooler, Kelly first saw and spoke with scientists from the disease-research institute when they stopped into the store to buy Diet Coke, quarts of milk, Nicorette gum, Tylenol . . . whatever it is that top-level, Army-affiliated virologists buy. Unlike your average young convenience-store clerk, Kelly herself had a strong early aptitude for science. During high-school summers she worked in a government institute of standards and measures. After her freshman year of college and each summer until graduation, she served as a laboratory assistant at the National Cancer Institute, which had a branch on the grounds of Fort Detrick. She finished a bachelor's degree in molecular biology and considered her options for grad school. Around the same time she read *The Hot Zone*, which had recently been published.

"I'm a *Hot Zone* kid," Warfield told me much later. She couldn't vouch for the book's scientific accuracy, she added, but its effect on her then was galvanic. She was inspired by one of the main characters, Nancy Jaax, an Army major and veterinary pathologist at USAMRIID, who had been part of the response team at the infected monkey house in Reston. Warfield herself hoped to return to Fort Detrick after graduate school and join USAMRIID as a scientist—if possible, to work on Ebola virus.

She looked for a doctoral program that would teach her virology and found a good one at Baylor College of Medicine, in Houston. An entire department at Baylor was devoted to viral research, with two dozen virologists, some of whom were quite eminent, though none dealt with such high-hazard pathogens as Ebola. Warfield found a place in the lab of a mentor there and began studying a group of gastrointestinal viruses, the rotaviruses, which cause diarrhea in humans. Her dissertation project looked at immune response against rotavirus infection in mice. That was intricate and significant work (rotaviruses kill a half million children around the world every year), though not especially dramatic. She got experience in using lab animals (especially mice) as models for human immune response to viral infections, and she learned a bit about making vaccines. In particular, she gained expertise in a line of vaccine development using viruslike particles (VLPs), rather than the more conventional approach, which uses live virus attenuated by laboratory-induced evolution. VLPs are essentially the outer shells of viruses, capable of inducing antibody production (immune readiness) but empty of functional innards, and therefore incapable of replicating or causing disease. VLPs seem to hold high promise for vaccines against viruses, such as Ebola, that might be too dangerous for live-virus vaccination.

It took some time for Kelly to achieve her dream, but not much, and she wasted none. With the doctorate finished, twenty-six-year-old Dr. Warfield began work at USAMRIID in June 2002, just days after her graduation in Houston. The Army's institute had hired her, in part, for her VLP skills. Immediately she enrolled in the Special Immunizations Program, a punishing series of shots and more shots required before a new person can be cleared to enter the BSL-3 labs. (BSL-3 comprises the laboratory suites in which researchers generally work on dangerous but curable diseases, many caused by bacteria, such as anthrax and plague. BSL-4 is reserved for work on pathogens such as Ebola, Marburg, the SARS virus, Machupo, and Nipah, for which there are neither vaccines nor treatments.) They vaccinated her against a whole list of unsavory things that she might or might not ever face in the lab—against Rift Valley fever, against Venezuelan equine encephalitis, against smallpox, and against anthrax—all within a year.

Some of these vaccines can make a person feel pretty sick. Anthrax, for Warfield, was a particular disfavorite. "Ooof, terrible!" she recalled, during our long conversation at her home, in a new suburb outside of Frederick. "That's a terrible vaccine." After all these challenges to her immune system, and possibly as a result, she suffered an attack of rheumatoid arthritis, which runs in her family. Rheumatoid arthritis is an immune dysfunction, and the medicine used to control it can potentially suppress normal immune responses. "So I wasn't allowed to get any vaccines anymore." Nonetheless, she was cleared to enter the BSL-3 suites, and then soon the BSL-4s. She began working with live Ebola virus.

Much of her effort went into the VLP research, though she also helped on other projects within her boss's lab. One involved testing a form of laboratory-created antibodies that might serve as a treatment against Ebola virus disease. These antibodies, developed by a private company in collaboration with USAMRIID, were designed to thwart the virus by tangling with a cellular protein involved in viral replication, not with the virus itself. It was a clever idea. Warfield again used mice as her test animals; she now had years of experience at handling and injecting them. For the experiment she infected fifty or sixty mice with Ebola virus and then, during the following days, gave them the experimental antibody treatment. Would they live, would they die? The mice were kept in clear plastic cages, like tall-sided pans, ten mice to a pan. Methodical procedures and constant attention are crucial to BSL-4 work, as Warfield well knew. Her methodical procedures for this experiment included filling a syringe full of antibody solution, enough for ten doses, and then injecting the ten mice from each pan with the same syringe, the same needle. It wasn't as though cross-infection was a concern, since they had already been dosed with the same batch of Ebola. Dosing multiple mice with a single syringe saved time, and time in a BSL-4 lab adds up toward stress and increased risk, because the physical circumstances are so difficult.

Picture those circumstances for Kelly Warfield. Customarily she worked in the BSL-4 suite known as AA-5, off a cinderblock corridor in the most secure wing of USAMRIID, behind three pressure-sealed doors and a Plexiglas window. She wore a blue vinyl protective suit (she and her colleagues simply called them "blue suits," not spacesuits or hazmats) with a fully enclosed hood, a clear face-shield, and a ventilation hookup. Attached to her hookup was a yellow hose, coiling down from the ceiling to bring filtered air. She wore rubber boots and two pairs of gloves—latex gloves beneath heavier canners gloves, sealed to her suit at the wrists with electrical tape. Even with canners gloves over latex, her hands were the most vulnerable part of her body; they couldn't be protected with vinyl because they had to be delicately dexterous.

Her workbench was a stainless steel cart, like a hospital cart, easy to clean, easy to move. If you didn't love the work, you wouldn't put yourself in this place.

She was alone in AA-5, under exactly those circumstances, at five thirty on the evening of February 11, 2004. She had come late to the day's tasks for the Ebola experiment because earlier hours had been filled with other demands. One pan of mice sat on her cart, along with a plastic beaker, a clipboard, and not much else in the way of materials and tools. It was the last pan of mice for the day. She filled a syringe and carefully injected nine mice, one after another—gripping each animal by the skin behind its neck, turning it belly up, inserting the needle into its abdomen deftly, quickly, adding no more discomfort than necessary to the life of each doomed and Ebola-ridden mouse. After each injection, she placed that mouse in the beaker, to keep the finished group apart from the others. One mouse to go. Maybe she was a little tired. Accidents happen. It was this very last mouse that caused the trouble. Just after being injected, it kicked away the needle, deflecting the point into the base of Kelly Warfield's left thumb.

The wound, if there was a wound, seemed to be only a very light graze. "At first, I didn't think that the needle went through the gloves," she told me. "It didn't hurt. Nothing hurt." Remaining calm by an act of discipline, she set the mouse back in its pan, put the syringe away, and then squeezed her hand. She could see blood emerging under the layers of glove. "So I knew I had stuck myself."

We were seated at her dinette table, on a mild September afternoon, as she talked me through the events of that February day. The house, which she shared with her Army-physician husband and her young son, was light and cheery with a lived-in feel; there were pieces of kid art on the refrigerator, a few toys lying around, a large green backyard, two half-poodle dogs, and a sign on the kitchen wall commanding: DO NOT ENTER WITHOUT WEARING VENTILATED SUIT. Today she was dressed in a red jacket and pearl earrings, not in blue vinyl.

She recalled her mind racing forward, from an immediate "Oh my God, I've done it" reaction to a sober consideration of just what she *had* done. She had not injected herself with live Ebola virus—or at least, not much. The syringe didn't carry Ebola virus; it carried antibodies, which would be harmless to anyone. But the needle had gone into ten Ebola-infected mice before going into her. If its point had picked up any particles of Ebola and brought them along, then she might have received a tiny dose. And she knew that a tiny dose could be enough. Quickly she unhooked her yellow hose and exited the BSL-4 suite, by way of the first of the pressurized doors, into an airlock space equipped with a chemical shower. There she showered out, dosing her blue-suit exterior with a virus-killing solution.

Then she pushed through the second door, to a locker-room area known as the Gray Side. She shed the boots, peeled off the blue suit and the gloves as fast as she could, leaving her clad only in medical scrubs. She used a wall phone to call two close friends, one of whom was Diane Negley, the BSL-4 suite supervisor. It was now suppertime or later, and Negley didn't answer at home, so Warfield left a chilling, desperate message on Negley's machine, the gist of which was: I've had an accident, stuck myself, please come back to work. The other friend, a co-worker named Lisa Hensley, who hadn't yet left the building, answered her call and said: "Start scrubbing. I'm on my way down." Warfield began scrubbing her hands with Betadine, rinsing with water and saline solution, scrubbing again. In her fervor she splashed water all over the floor. Hensley arrived quickly, joined her in the Gray Side, and started making calls to alert other people, including those in the Medical Division who handled accidents, while Warfield continued the Betadine scrub. After five or ten minutes, feeling she had done what she could on the wound site, Warfield stripped out of her medical scrubs, took a soap-and-water shower, and dressed. Hensley did likewise. But when they tried to exit the Gray Side, that pressure-sealed door wouldn't open. Its electronic lock didn't respond to their badges. Warfield, full of adrenaline, scared, with no luxury of being patient, busted open the door on manual override and alarms started ringing in other parts of the building.

Word had spread fast through the institute and, by now, a small crowd had gathered in the corridor. Warfield passed amid their stares and their questions, headed for the Medical Division. There she was ushered into a small room, questioned about her accident by the doctor on duty, a civilian woman, and given a "physical exam," through the whole course of which the doctor never touched her. "It was like she was afraid that I already had Ebola," Warfield recalled. The incubation period for Ebola virus is measured in days, not hours or minutes. It takes at least two days and usually more than a week for the virus to establish itself, replicate abundantly, and make a person symptomatic or infectious. But the civilian doctor didn't seem to know that, or to care. "She acted like I was a leper already." That doctor went off to confer with others, after which the head of the Medical Division took Warfield into his office, sat her down, and gently told her the recommended next step. They wanted to put her in the Slammer.

The Slammer at USAMRIID is a medical containment suite, designed for care of a person infected with any dangerous pathogen and—equally—for protecting against the spread of that infection to others. It consists of two hospital-style rooms set behind more pressure-sealed doors and another chemical shower. Earlier on the day of our conversation, having gotten me clearance for a tour of USAMRIID, Warfield had shown me through the Slammer, explaining its features with a trace of mordant pride. On the outside, a wide main door is labeled: CONTAINMENT ROOM. AUTHORIZED PERSONNEL ONLY. That's door number 537 within USAMRIID's labyrinthine corridors. It's the door through which a new patient enters the suite and, if things go well, through which the same patient eventually walks out. If things don't go well, the patient exits under other circumstances, not walking and not via door 537. All other human traffic—the flow of medical caregivers and faithful, intrepid friends—must pass through a smaller door into a change room, where piles of scrub suits sit folded and ready on shelves, and then through a pressurized steel door into an airlock shower. On the other side of the shower stall is another steel door. The two pressurized steel doors are never both open at once. So long as the patient shows no signs of infection, approved visitors are admitted to the Slammer wearing scrubs, gowns, masks, and gloves. If the patient proves to be infected, the suite becomes an active BSL-4 zone, in which doctors and nursing staff (no visitors now) must wear full blue suits. In that situation, the medical people shower thoroughly on the way out, leaving their scrub clothing behind in a bag to be autoclaved.

Warfield led me. We could pass through the shower stall in street clothes because the containment suite was unoccupied. When she slammed the first steel door behind us, triggering pressurization, I heard a *voosh* and felt the change in my ears. She said: "There's why it's called the Slammer."

She had entered the suite around noon on February 12, 2004, the day following her accident, after having drawn up a will and an advance directive (stipulating end-of-life medical decisions) with help from an Army lawyer. Her husband was in Texas for advanced military training and she had apprised him of the situation by phone. In fact, she had stayed on the phone with him much of the previous night, helped through the hours of terror and dread by his long-distance support. At some point she told him: "If I get sick, please *please* give me a lot of morphine. I've seen this disease"—she had watched it kill monkeys in the lab, though never a human—"and I know it *hurts*." On the first weekend, he managed to fly up from Texas and they spent Valentine's Day in the suite holding hands through his latex gloves. There was no kissing through his mask.

The incubation period for Ebola virus disease, as I've mentioned, is reckoned to be at least two days; it can be longer than three weeks. Individual case histories differ, of course, but at that time twenty-one days seemed to be the outer limit. Expert opinion held that, if an exposed person hasn't shown the disease within that length of time, she wouldn't. Kelly Warfield was therefore sentenced to twenty-one days in the Slammer. "It was like prison," she told me. Then she amended her statement: "It's like prison *and* you're gonna die."

Another difference from prison is that there were more blood tests. Each morning her friend Diane Negley, who happened to be a certified phlebotomist and who knew enough about Ebola to be cognizant of the risk to herself, tapped a vein and took away some of Warfield's blood. In exchange, she brought a donut and a latte. Negley's morning visit was the highlight of Warfield's day. During the first week or so, Negley took fifty milliliters of blood daily, a sizable volume (more than three tablespoons) that allowed for multiple tests plus a bit extra to put in frozen storage. One test, using the PCR (polymerase chain reaction) technique that's familiar to all molecular biologists, looked for sections of Ebola RNA (the virus's genetic molecule, equivalent to human DNA) in her blood. That test, which can ring a loud alarm but is sometimes unreliable, delivering a false positive, was routinely performed twice on each sample. Another test screened for interferon, the presence of which might signal a viral infection of any sort. Still another test targeted changes in blood coagulation, for an early alert in case of disseminated intravascular coagulation, the catastrophic clotting phenomenon that makes blood ooze out where it shouldn't. Warfield encouraged the medical people to take all the blood they desired. She recalled telling them: "If I die, I want you to learn everything you can about me"—everything they could about Ebola virus disease, she meant. "Store every sample. Analyze everything you can. Please *please* take something away from this if I die. I want you to learn." She told her family the same: If the worst happens, let them autopsy me. Let them salvage all possible information.

If she did die, Warfield knew, her body wouldn't come out of the Slammer through door 537. After autopsy, it would come through the autoclave chute, a sterilizing cooker, which would leave nothing her loved ones would want to see in an open coffin.

All her test results during the first week were normal and reassuring—with a single exception. The second PCR test from one day's sample came back positive. It said she had Ebola virus in her blood.

It was wrong. The provisional result gave Warfield a fright but that mistake was soon corrected by further testing. Woops, no, sorry. Never mind.

Another kerfuffle arose when USAMRIID's leadership realized that Warfield suffered rheumatoid arthritis, the medications for which might have suppressed her immune system. "That became this huge controversy," she told me. Certain honchos of the institute's top leadership acted surprised and angry, although the condition was clearly on file in her medical records. "They had all these teleconferences with all these experts. Everybody wanted to know why someone that was immunocompromised was working in the BSL-4 suites." There was in fact no evidence that her immune system wasn't working fine. The commander of USAMRIID never made a personal visit to see her in the Slammer, not even through the glass, but he sent her an email announcing that he was suspending her access to BSL-4 labs and impounding her badge. It was a "slap in the face," added onto her other miseries and worries, Warfield said.

After more than two weeks of vampiric blood draws and reassuring tests, Warfield began feeling guardedly confident she wouldn't die of Ebola. She was weak and weary, her veins were weary too, so she asked that the blood sampling be reduced to a daily minimum. She got another unsettling jolt one evening as she undressed, discovering red spots on her arm and wondering whether they might herald the start of Ebola's characteristic rash. She had seen similar spots on lab-infected monkeys. That night she lay awake, obsessing about the spots, but they turned out to be nothing. She had Ambien to help her sleep. She had a stationary bike in case she wanted exercise. She had TV and Internet and a phone. As the weeks passed, the terrifying element of her situation faded slowly beneath the good news and the tedium.

She stayed sane with help from her mother and a few close friends (who could visit her often), her husband (who couldn't), her father (who remained off the visitor list so he could look after her son, in case everyone else got infected and quarantined and then died), and a certain amount of nervous laughter. Her son, whose name is Christian, was just three at the time and barred by age regulations from entering USAMRIID. Warfield judged he was too young, in any case, to be burdened with knowing exactly what was going on; she and her husband explained to Christian simply that mom would be absent for three weeks doing "special work." She was given a video linkup, a sort of Slammer Cam, through which she could see and talk with her loved ones on the outside. Hi, it's me, Kelly, live from Ebolaville, how was your day? Diane Negley, besides supplying the morning donut and coffee, heroically smuggled in one beer every Friday night. Food was a problem at first, there being no cafeteria at USAMRIID, until the Army realized it had funds that could be spent on supplying a patient in the Slammer with carryout. After that, Warfield had her choice each evening among Frederick's best: Chinese, Mexican, pizza. And she could share with her visiting friends, such as Negley, who would sit in the blind spot beneath the security camera, flip up her face shield, and eat. These high-carb consolations led Warfield and her pals to invent a game: "*Ebola Makes You . . .*" and then fill in the blank. Ebola makes you fat. Ebola makes you silly. Ebola makes you diabetic from too much chocolate ice cream. Ebola makes you appreciate little joys and smiles in the moment.

On the morning of March 3, 2004, door 537 opened and Kelly Warfield walked out of the Slammer. Her mother and (by special exemption) Christian were in the waiting room down the corridor. She took her son home. That afternoon she returned to USAMRIID, where her friends and colleagues threw her a coming-out party with food, testimonials, and balloons. Several months later, after a period of suspended access, a battery of tests on her immune system, a somewhat humiliating regimen of retraining and supervision, and a bit of persistent struggle, she regained her clearance for the BSL-4 suites. She could return to tickling the tail of the dragon that might have killed her.

Did you ever consider *not* going back to Ebola? I asked.

"No," she said.

Why do you love this work so much?

"I don't know," she said, and began to ruminate. "I mean, *why* Ebola? It only kills maybe a couple hundred people a year." That is, it hadn't been a disease of massive global significance, not at the time of this conversation, in 2008 (although, amid the frightful events of 2014, it has certainly captured global attention, while increasing its death toll and breadth of impact alarmingly). Apart from the numbers, she could cite its attractions in scientific terms. She took deep interest, for instance, in the fact that such a simple organism can be so potently lethal. It contains only a tiny genome, enough to construct just ten proteins, which account for the entire structure, function, and self-replicating capacity of the thing. (A herpesvirus, by contrast, carries about ten times more genetic complexity.) Despite the minuscule genome, Ebola virus is ferocious. It can kill a person in seven days. "How can something that is so small and so simple just be so darn dangerous?" Warfield posed the question and I waited. "That's just really fascinating to me."

Her son Christian, grown to a handsome first-grader, at this point arrived home from school. Kelly Warfield had given me most of her day and now there was time for just one more question. Although she is a molecular biologist, not an ecologist, I mentioned those two unsolved mysteries of Ebola's life in the wild: the reservoir host and the spillover mechanism.

Yes, very intriguing also, she agreed. "It pops up and kills a bunch of people, and before you can get there and figure anything out, it's gone."

It disappears back into the Congo forest, I said.

"It disappears," she agreed. "Yeah. Where did it come from and where did it go?" But that was out of her area.

12

Think of a BSL-4 laboratory—not necessarily AA-5 at USAMRIID but any among a handful around the world in which this virus is studied. Think of the proximity, the orderliness, and the certitude. Ebola virus is in these mice, replicating, flooding their bloodstreams. Ebola virus is in that tube, frozen solid. Ebola virus is in the Petri dish, forming plaques among human cells. Ebola virus is in the syringe; beware its needle. Now think of a forest in northeastern Gabon, just west of the upper Ivindo River. Ebola virus is everywhere and nowhere. Ebola virus is present but unaccounted for. Ebola virus is near, probably, but no one can tell you which insect or mammal or bird or plant is its secret repository. Ebola virus is not in *your* habitat. You are in *its*.

That's how Mike Fay and I felt as we hiked through the Minkébé forest in July 2000. Six days after my helicopter fly-in we left the inselbergs area, trudging southwest on Fay's compass line through a jungle of great trees, thorny vines interwoven into torturous thickets, small streams and ponds, low ridges between the stream drainages, mud-bordered swamps dense with thorny vegetation, fallen fruits as big as bocce balls, driver ants crossing our path, groups of monkeys overhead, forest elephants in abundance, leopards, almost no signs of human visitation, and roughly a trillion cheeping frogs. The reservoir host of Ebola virus was there too, presumably, but we couldn't have recognized it as that if we'd looked it in the face. We could only take sensible precautions.

On the eleventh day of walking, one of Fay's forest crewmen spotted a crested mona monkey on the forest floor, a youngster, alive but near death, with blood dripping from its nostrils. Possibly it had missed its grip in a high tree and suffered a fatal fall. Or . . . maybe it was infected with something, such as Ebola, and came down to die. Under standing instructions from Fay, the crewman didn't touch it. Fay's crew of hardworking Bantus and

Pygmies always hungered after wild meat for the evening pot, but he forbade hunting on conservation grounds—and during this stretch through Minkébé he had commanded his cook even more sternly: Do *not* feed us anything found dead on the ground. That night we ate another brownish stew, concocted from the usual freeze-dried meats and canned sauces, served over instant mashed potatoes. The dying monkey, I fervently hoped, had been left behind.

One night later, at the campfire after dinner, Fay helped me tease some direct testimony from Sophiano Etouck, the shier of the two survivors from Mayibout 2. I had heard the whole story—including the part about Sophiano's personal losses—from the voluble Thony M'Both, but Sophiano himself, burly, diffident, had never spoken up. Now finally he did. The sentences were diced cruelly by his stutter, which sometimes brought him to what seemed an impassable halt; but Sophiano pushed on, and between blockages his words came quickly.

He had been traveling to one of the gold camps. Farther upriver. And stopped in Mayibout 2 to stay with family. That night one of his nieces said she was feeling bad. Malaria, everyone thought. A routine thing. The next morning, it got worse. Then other people too. They vomited, they had diarrhea. Started dying. I lost six, Sophiano said. Thony had gotten the number right but was a little confused about the identities. An uncle, a brother, a widowed sister-in-law. Her three daughters. The men in white suits, they came to take charge. One of them, a Zairian, had seen the disease before. At Kikwit. Twenty doctors had died there at Kikwit, the Zairian told us. They told us, this thing is very infectious. If a fly lands on you after having touched one of the corpses, they said, you will die. But I held one of my nieces in my arms. She had a tube in her wrist, an IV drip. It got clogged, backed up. Her hand swelled. And then with a pop her blood sprayed all over my chest, Sophiano said. But I didn't get sick. You've got to take the remedy, the doctors told me. You've got to stay here twenty-one days under quarantine. I thought, the hell with that. I didn't take the remedy. After my family people had been buried, I left

Mayibout 2. I went to Libreville and stayed with another sister, hiding, Sophiano confessed. Because I was afraid the doctors would hassle me, he said.

This was our last evening in the forest before a resupply rendezvous four or five miles onward, at a point where Fay's pre-plotted line of march crossed a road. That road led eastward to Makokou. Some of Fay's crew would leave him there. They were exhausted, spent, fed up. Others would stay with him because, though also exhausted, they needed the work badly, or because it was better than gold mining, or because those reasons supplemented another: the sheer fascination of being involved with an enterprise so sublimely crazed and challenging. Another half year of hard walking across forests and swamps lay between them and Fay's end point, the Atlantic Ocean.

Sophiano would stay. He had been through worse.

13

The identity of Ebola's reservoir host (or hosts) remains unknown, as of this writing, although suspects have been implicated. Several different groups of researchers have explored the question. The most authoritative, most advantageously placed, and most persistent of them is the team led by Eric M. Leroy, of CIRMF, in Franceville, Gabon. As mentioned earlier, Leroy was one of the visiting doctors dressed in mystifying white suits who took part in the response effort at Mayibout 2. Although he and his colleagues may not have saved many (or *any*, as remembered by Thony M'Both) of the Mayibout patients from death, that outbreak was transformative for Leroy himself. He trained as an immunologist as well as a veterinarian and a virologist, and until 1996 he studied the effects of another kind of virus on the immune systems of mandrills. Mandrills are large, baboonlike monkeys with red noses, puffy blue facial ridges, and contorted expressions, all of which give them the look of angry, dark clowns. Leroy was also curious about the immune physiology of bats. Then came Mayibout 2 and Ebola.

"It is a little bit like a fate," Leroy told me when I visited him in Franceville.

Back at CIRMF after Mayibout 2, he explored Ebola further in his lab. He and a colleague, like him an immunologist, investigated some molecular signals in blood specimens taken during the outbreak. They found evidence suggesting that the medical outcome for an individual patient—to survive and recover, or to die—might be related not to the size of the infectious dose of Ebola virus but to whether the patient's blood cells produced antibodies promptly in response to infection. If they didn't, why not? Was it because the virus itself somehow quickly decommissioned their immune systems, interrupting the normal sequence of molecular interactions involved in antibody production? Does the virus kill people (as is now widely supposed) by creating immune dysfunction before overwhelming them with viral replication, which then inflicts further devastating effects? Leroy and his immunologist colleague, with a group of additional coauthors, published this study in 1999, after which he became interested in other dimensions of Ebola: its ecology and its evolutionary history.

The ecology of Ebola virus encompasses the reservoir question: Where does it hide between outbreaks? Another ecological matter is spillover: By what route, and under what circumstances, does the virus pass from its reservoir into other animals, such as apes and humans? To ask those questions is one thing; to get data that might help answer them is more tricky. How does a scientist study the ecology of such an elusive pathogen? Leroy and his team went into the forest, near locations where Ebola-infected gorilla or chimp carcasses had recently been found, and began trapping animals wholesale. They were groping. Ebola might abide in one of these creatures—but which one?

In the course of several expeditions between 2001 and 2003,

into Ebola-stricken areas of Gabon and the Republic of the Congo, Leroy's group caught, killed, dissected, and took samples of blood and internal organs from more than a thousand animals. Their harvest included 222 birds of various species, 129 small terrestrial mammals (shrews and rodents), and 679 bats. Back at the lab in Franceville, they tested the samples for traces of Ebola using two different methods. One method was designed to detect Ebola-specific antibodies, which would be present in animals that had responded to infection. The other method used PCR (as it had been used on Kelly Warfield) to screen for fragments of Ebola's genetic material. Having looked so concertedly at the bat fauna, which accounted for two-thirds of his total collections, Leroy found something: evidence of Ebola virus infection in bats of three species.

These three were all fruit bats, relatively big and ponderous. One of them, the hammer-headed bat (*Hypsignathus monstrosus*), is the largest bat in Africa, as big as a crow. People hunt it for food. But in this case the evidence linking bats and virus, though significant, wasn't definitive. Sixteen bats (including four hammer-headed) had antibodies. Thirteen bats (again including some hammer-headed) had bits of the genome of Ebola virus, detectable by PCR. That amounted to twenty-nine individuals, representing a small fraction of the entire sample. And the results among even those twenty-nine seemed ambiguous, in that no individual bat tested positive by both methods. The sixteen bats with antibodies contained no Ebola RNA, and vice versa. Furthermore, Leroy and his team did not find live Ebola virus in a single bat—nor in any of the other animals they opened.

Ambiguous or not, these results seemed dramatic when they appeared in a paper by Leroy and his colleagues in late 2005. It was a brief communication, barely more than a page, but published by *Nature*, one of the world's most august scientific journals. The headline ran: FRUIT BATS AS RESERVOIRS OF EBOLA VIRUS. The text itself, more carefully tentative, said that bats of three species "may be acting as a reservoir" of the virus. Some experts reacted as though the question were now virtually settled, others reserved judgment. "The only thing missing to be sure that bats are the reservoir," Leroy told me, during our conversation ten months later, "is virus isolation. Live virus from bats." That was 2006. It still hasn't happened, so far as the world knows, though not for lack of effort on his part. "We continue to catch bats—to try to isolate the virus from their organs," he said.

But the reservoir question, Leroy emphasized, was only one aspect of Ebola that engaged him. Using the methods of molecular genetics, he was also studying its phylogeny—the ancestry and evolutionary history of the whole filovirus lineage, including Marburg virus and the various ebolaviruses. He wanted to learn too about the natural cycle of the virus, how it replicates within its reservoir (or reservoirs) and maintains itself in those populations. Finally, knowing something about the natural cycle would help in discovering how the virus is transmitted to humans: the spillover moment. Does that transmission somehow occur directly (for instance, by people eating bats), or through an intermediate host? "We don't know if there's direct transmission from bats to humans," he said. "We only know there is direct transmission from dead great apes to humans." Understanding the dynamics of transmission—including seasonal factors, the geographical pattern of outbreaks, and the circumstances that bring reservoir animals or their droppings into contact with apes or humans—might give public health authorities a chance to predict and even prevent some outbreaks. But there exists a grim circularity: Gathering more data requires more outbreaks.

Ebola is difficult to study, Leroy explained, because of the character of the virus. It strikes rarely, it progresses quickly through the course of infection, it kills or it doesn't kill within just a few days, it affects only dozens or hundreds of people in each outbreak, and those people generally live in remote areas (again, he was right on both points until 2014), far from research hospitals and medical institutes—far even from his institute, CIRMF. (It takes about two days to travel, by road and river, from Franceville to Mayibout 2.) Then the outbreak exhausts itself locally, coming to a dead end, or is successfully stanched by intervention. The virus disappears like a band of jungle guerrillas. "There is nothing to do," Leroy said, expressing the momentary perplexity of an otherwise patient man. He meant, nothing to do except keep trying, keep working, keep sampling from the forest, keep responding to outbreaks as they occur. No one can predict when and where Ebola virus will next spill. "The virus seems to decide for itself."

14

The geographical pattern of Ebola outbreaks among humans is controversial. Everyone knows what that pattern looks like but experts dispute what it means. The dispute involves Ebola virus in particular, the one among those five ebolaviruses that has emerged most frequently, in multiple locations across Africa, and therefore cries most loudly for explanation. From its first known appearance to the present, from Yambuku (1976) to Tandala (1977) to the upper Ivindo River gold camps (1994) to Kikwit (1995) to Mayibout 2 (1996) to Booué (later 1996) to the northern border region between Gabon and the Republic of the Congo (2001–2002) to the Mbomo area (2002–2003) to its recurrence at Mbomo (2005), and then to its two appearances near the Kasai River in what's now the Democratic Republic of the Congo (2007–2009), and most recently to Guinea, Sierra Leone, Liberia, and Nigeria (2013–2014), Ebola virus has seemingly hopscotched its way around Central and West Africa. What's going on? Is that pattern random or does it have causes? If it has causes, what are they?

Two schools of thought have arisen. I think of them as the wave school and the particle school—my little parody of the classic wave-or-particle conundrum about the nature of light. Back in the seventeenth century, as your keen memory for high-school physics will tell you, Christiaan Huygens proposed that light consists of waves, whereas Isaac Newton argued that light is particulate. They each had some experimental grounds for believing as they did. It took quantum mechanics, more than two centuries later, to explain that wave-versus-particle is not a resolvable dichotomy but an ineffable duality, or at least an artifact of the limitations of different modes of observing.

The particle view of Ebola sees it as a relatively old and ubiquitous virus in Central African forests, and each human outbreak as an independent event, primarily explicable by an immediate cause. For instance: Somebody scavenges an infected chimpanzee carcass; the carcass is infected because the chimp itself scavenged a piece of fruit previously gnawed by a reservoir host. The subsequent outbreak among humans results from a local, accidental event, each outbreak therefore representing a particle, discrete from others. Eric Leroy is the leading proponent of this view. "I think the virus is present all the time, within reservoir species," he told me. "And sometimes there is transmission from reservoir species to other species."

The wave view suggests that Ebola has *not* been present throughout Central Africa for a long time—that, on the contrary, it's a rather new virus, descended from some viral ancestor, perhaps in the Yambuku area, and come lately to other sites where it has emerged. The local outbreaks are not independent events, but connected as part of a wave phenomenon. The virus has been expanding its range within recent decades, infecting new populations of reservoir in new places. Each outbreak, by this view, represents a local event primarily explicable by a larger cause—the arrival of the wave. The main proponent of the wave idea is Peter D. Walsh, an American ecologist who has worked often in Central Africa and specializes in mathematical theory about ecological facts.

"I think it's spreading from host to host in a reservoir host," Walsh said, when I asked him to explain where the virus was traveling and how. This was another conversation in Libreville, a teeming Gabonese city with pockets of quietude, through which all Ebola researchers eventually pass. "Probably a reservoir host that's got large population sizes and doesn't move very much. At least, it doesn't transmit the virus very far." Walsh didn't claim to know the identity of that reservoir, but it had to be some animal that's abundant and relatively sedentary. A rodent? A small bird? A nonmigrating bat?

The evidence on each side of this dichotomy is varied and intriguing, though inconclusive. One form of that evidence is the genetic differences among strains of Ebola virus as they have been found, or left traces of themselves, in human victims, gorillas, and other animals sampled at different times and places. Ebola virus in general seems to mutate at a rate comparable to other RNA viruses (which means relatively quickly), and the amount of variation detectable between one strain of Ebola virus and another can be a very important clue about their origins in space and time. Peter Walsh, working with two coauthors on a paper published in 2005, combined such genetic data with geographical analysis to suggest that all known variants of Ebola virus descended from an ancestor closely resembling the Yambuku virus of 1976.

Walsh's collaborators were Leslie Real, a highly respected disease ecologist and theoretician at Emory University, and a bright younger colleague named Roman Biek. Together they presented maps, graphs, and family trees illustrating strong correlations among three kinds of distance: distance in miles from Yambuku, distance in time from that 1976 event, and distance in genetic differences from the Yambuku-like common ancestor. "Taken together, our results clearly point to the conclusion that [Ebola virus] has gradually spread across central Africa from an origin near Yambuku in the mid-1970s," they wrote. Their headline, stating the thesis plainly, was WAVE-LIKE SPREAD OF EBOLA ZAIRE. It may or may not be a new pathogen—at least, new in these places. (Other evidence, published more recently, suggests that filoviruses may be millions of years old.) But maybe something happened, and happened rather recently, to reshape the virus and unleash it upon humans and apes. "Under this scenario, the distinct phylogenetic tree structure, the strong correlation between outbreak date and distance from Yambuku, and the correlation between genetic and geographic distances can be interpreted as the outcome of a consistently moving wave of [Ebola virus] infection." One consequence of the moving wave, they argued, is massive mortality among the apes. Some regional populations have been virtually exterminated—such as the gorillas of the Minkébé forest, of the Lossi Gorilla Sanctuary, of the area around Moba Bai—because Ebola hit them like a tsunami.

So much for the wave hypothesis. The particle hypothesis embraces much of the same data, construed differently, to arrive at a vision of independent spillovers, not a traveling wave. Eric Leroy's group also collected more data, including samples of muscle and bone from gorillas, chimps, and duikers found dead near human outbreak sites. In some of the carcasses (especially the gorillas), they detected evidence of Ebola virus infection, with small but significant genetic differences in the virus among individual animals. Likewise they looked at a number of human samples, from the outbreaks in Gabon and the Republic of the Congo during 2001–2003, and identified eight different viral variants. (These were lesser degrees of difference than the gaps among the five ebolaviruses.) Such distinct viruses, they proposed, should be understood in the context that their genetic character is relatively stable. The differences among variants suggest long isolation in separate locales, not a rolling wave of newly arrived, rather uniform virus. "Thus, Ebola outbreaks probably do not occur as a single outbreak spreading throughout the Congo basin as others have proposed," Leroy's team wrote, alluding pointedly to Walsh's hypothesis, "but are due to multiple episodic infection of great apes from the reservoir."

This apparent contradiction between Leroy's particle hypothesis and Walsh's wave hypothesis reflects an argument at cross-purposes, I think. The confusion may have arisen from back-channel communications and a certain sense of competition as much as from ambiguity in their published papers. What Walsh suggested—to recapitulate in simplest form—is a wave of

Ebola virus sweeping across Central Africa by newly infecting some reservoir host or hosts. From its recent establishment in the host, according to Walsh, the virus spilled over, here and there, into ape and human populations. The result of that process is manifest as a sequence of human outbreaks coinciding with clusters of dead chimps and gorillas—*almost* as though the virus were sweeping through ape populations across Central Africa. Walsh insisted during our Libreville chat, though, that he had never proposed a continental wave of dying gorillas, one group infecting another. His wave of Ebola, he explained, has been traveling mainly through the reservoir populations, not through the apes. Ape deaths have been numerous and widespread, yes, and to some degree amplified by ape-to-ape contagion, but the larger pattern reflects progressive viral establishment in some other group of animals, still unidentified, with which apes frequently come into contact. Leroy, on the other hand, has presented his particle hypothesis of "multiple independent introductions" as a diametric alternative not to Walsh's idea as here stated but to the notion of a continuous wave among the apes.

In other words, one has cried: *Apples!* The other has replied: *Not oranges, no!* Either might be right, or not, but in any case their arguments don't quite meet nose to nose.

So . . . is light a wave or a particle? The coy, modern, quantum-mechanical answer is *yes*. And is Peter Walsh correct about Ebola virus or is Eric Leroy? The best answer again may be *yes*. Walsh and Leroy eventually coauthored a paper, along with Roman Biek and Les Real as deft reconcilers, offering a logical amalgam of their respective views on the family tree of Ebola virus variants (all descended from Yambuku) and of the hammer-headed bat and those two other kinds of bats as (relatively new) reservoir hosts. But even that paper left certain questions unanswered, including this one: If the bats have just recently become infected with Ebola virus, why don't they suffer symptoms?

The four coauthors did agree on a couple of other basic points. First, fruit bats might be reservoirs of Ebola virus but not necessarily the *only* reservoirs. Maybe another animal is involved—a more ancient reservoir, long since adapted to the virus. (If so, where is *that* creature hiding?) Second, they agreed that too many people have died of Ebola virus disease, but not nearly so many people as gorillas.

15

Why do we share these diseases with wildlife, and why do they seem to be emerging ever more frequently—Ebola one year, bird flu another year, then SARS out of China, then something called MERS from the Arabian peninsula, then something else, and then Ebola again?

"The key is connectivity," Jon Epstein told me. He's a veterinary disease ecologist, based in New York and traveling the world for EcoHealth Alliance, the same organization that employs Billy Karesh. Epstein and I were in Bangladesh at the time, where he was trapping giant fruit bats in search of the virus called Nipah, little known outside of Asia but roughly as lethal as Ebola. "The key is to understand how animals and people are interconnected." You can't look at a new bug or a reservoir host as though they exist in a vacuum, Epstein said. It's a matter of contact with humans, interaction, opportunity. "Therein lies the risk of spillover."

Repeatedly over the next half hour he returned to the word "opportunity." It kept knocking. "A lot of these viruses, a lot of these pathogens that come out of wildlife into domestic animals or people, have existed in wild animals for a very long time," he said. They don't necessarily cause any disease. They have coevolved with their natural hosts over millions of years. They have reached some sort of accommodation, replicating slowly but steadily, passing unobtrusively through the host population, enjoying long-term security—and eschewing short-term success in the form of maximal replication within each host individual.

It's a strategy that works. But when we humans disturb the accommodation—when we encroach upon the host populations, hunting them for meat, dragging or pushing them out of their ecosystems, disrupting or destroying those ecosystems—our action increases the level of risk. "It increases the opportunity for these pathogens to jump from their natural host into a new host," he said. The new host might be any animal (the chimpanzee or the gorilla, for instance) but often it's humans, because we are present so intrusively and abundantly. We offer a wealth of opportunity.

"Sometimes nothing happens," Epstein said. A leap is made but the microbe remains benign in its new host, as it was in the old one. (There is a thing called simian foamy virus, for instance, which gets into people from monkeys but causes no known human disease.) In other cases, the result is very severe disease for a limited number of people, after which the pathogen comes to a dead end. (Hendra virus in Australia kills about half the people it infects, but it infects few, and it hasn't passed from human to human.) In still other cases, the pathogen achieves great and far-reaching success in its new host. It finds itself well enough suited to get a foothold; it makes itself still better suited by adapting. It evolves, it flourishes, it continues. The history of HIV is the story of a leaping virus that might have come to a dead end but didn't.

Yes, HIV is a vivid example, I agreed. But is there any particular reason why other viruses shouldn't have the same potential? For instance, Nipah?

"No reason at all. There's no reason at all," Epstein said. "A lot of what determines whether a pathogen becomes successful in a new host, I think, is odds. Chance, to a large degree." Many of these new pathogens are RNA viruses, carrying their genomes on a single-stranded molecule, not the double-stranded molecule that is DNA. A single-stranded genome tends to yield more mistakes during replication, meaning a high rate of mutation. With their high rates of mutation and replication, RNA viruses are very adaptable, Epstein reminded me, and every spillover presents a new opportunity to adapt and take hold. We'll probably never know how often that occurs—how many animal viruses spill into people inconspicuously. Some of those viruses cause no disease, or they cause a new disease that—in some parts of the world, because health care is marginal—gets mistaken for an old disease. "The point being," he said, "that the more opportunity viruses have to jump hosts, the more opportunity they have to mutate when they encounter new immune systems." Their mutations are random but frequent, combining nucleotide bases (the coding elements of RNA and DNA) in myriad new ways. "And, sooner or later, one of these viruses has the right combination to adapt to its new host."

This point about opportunity is a crucial idea, more subtle than it might seem. I had heard it from a few other disease scientists. It's crucial because it captures the randomness of the whole situation, without which we might romanticize the phenomena of emerging diseases, deluding ourselves that these new viruses attack humans with some sort of purposefulness. (Loose talk about "the revenge of the rainforest" is one form of such romanticizing. That's a nice metaphor, granted, but shouldn't be taken too seriously.) Epstein was talking, in an understated way, about the two distinct but interconnected dimensions of zoonotic transfer: ecology and evolution. Habitat disturbance, bushmeat hunting, the exposure of humans to unfamiliar viruses that lurk in animal hosts—that's ecology. Those things happen *between* humans and other kinds of organisms, and are viewed in the moment. Rates of replication and mutation of an RNA virus, differential success for different strains of the virus, adaptation of the virus to a new host—that's evolution. It happens *within* a population of some organism, as the population responds to its environment over time. Among the most important things to remember about evolution—and about its primary mechanism, natural selection, as limned by Darwin and his successors—is that it doesn't have purposes. It only has results. To believe otherwise is to embrace a teleological fallacy that carries emotive appeal but misleads. This is what Jon Epstein was getting at. Don't imagine that these viruses have a deliberate strategy, he said. Don't think that they bear some malign onus against humans. "It's all about opportunity." They don't come after us. In one way or another, we go to them. We offer them opportunity to infect us when we mess with their reservoir hosts, whatever those creatures may be.

16

The subject of reservoir hosts has continued to intrigue and puzzle infectious disease scientists generally, not merely veterinary ecologists such as Epstein, and not merely those who study Ebola virus. In their pursuit of understanding, they look for patterns. They note that some kinds of animals are more deeply implicated than others as reservoirs of the zoonotic viruses that jump into humans.

Hantaviruses jump from rodents. Lassa virus too jumps from rodents. Yellow fever virus jumps from monkeys. Monkeypox, despite its name, seems to jump mainly from squirrels. Herpes B jumps from macaques. The influenzas jump from wild birds into domestic poultry and then into people, sometimes after a transformative stopover in pigs. Measles may originally have jumped into us from domesticated sheep and goats. HIV-1 has jumped our way from chimpanzees. So there's a certain diversity of origins. But a large fraction of all the scary new viruses known to be zoonoses, and for which reservoir hosts have been identified, come jumping at us from bats.

Hendra virus in Australia: from bats. Marburg in Africa: from bats. SARS coronavirus, which came out of China in 2003: from bats. Nipah virus in Malaysia, 1998, and then again in Bangladesh, 2001: from bats. Rabies, when it jumps into people, comes usually from domestic dogs—because mad dogs get more opportunities than mad wildlife to sink their teeth into humans—but bats are among its chief reservoirs. Duvenhage, a rabies cousin, jumps to humans from bats. Kyasanur Forest virus is vectored by ticks, which carry it to people from several kinds of wildlife, including bats. Menangle virus: from bats. Tioman virus: from bats. Melaka virus: from bats. Australian bat lyssavirus, it may not surprise you to learn, has its reservoir in Australian bats. And the evidence on Ebola virus, though not definitive, as I've mentioned, suggests that it too very possibly comes from bats.

Why bats? Before even considering that question, I suppose it's necessary to remind ourselves that bats are wondrous, valuable, and necessary animals, filling a variety of crucial roles—as insectivores, pollinators, seed dispersers, and otherwise—within the ecosystems of which they are integral parts. But what is it about the chiropteran order of mammals (or about our relations with them) that accounts for their seemingly inordinate role also as reservoir hosts of novel and nasty zoonotic viruses? I've put that question to emerging-disease experts around the world. One of them was Charles H. Calisher, an eminent virologist recently retired as professor of microbiology at Colorado State University.

Calisher came out of the Georgetown School of Medicine with a PhD in microbiology in 1964. He made his bones doing classic lab-table virology, which meant growing live viruses, passaging them experimentally through mice and cell cultures, looking at them in electron micrographs, figuring out where to place them on the viral family tree—the kind of work that Karl Johnson had done on Machupo virus, and that traced back before Johnson to other infectious-disease pioneers. Calisher's career included a long stretch at the CDC as well as academic appointments, during which he had focused on arthropod-borne viruses (such as West Nile, dengue, and La Crosse virus, all carried by mosquitoes) and rodent-borne viruses (notably the hantaviruses). As a scientist who studied viruses in their vectors and in their reservoirs for more than four decades, but with no particular attention to chiropterans, he had eventually found himself wanting to know: Why are so many of these new things emerging from *bats*?

Charlie Calisher is a smallish man with a dangerous twinkle, famed throughout the profession for his depth of knowledge, his caustic humor, his disdain for pomposity, his brusque manner, and (if you happen to get past those crusts) his big, affable heart. He insisted on buying me lunch, at a favorite Vietnamese restaurant in Fort Collins, before we got down to serious talk. He wore a fisherman's sweater, chinos, and hiking boots. After the meal I followed his red pickup truck back to a CSU laboratory compound, where he still had a few projects going. He pulled a flat-sided flask from an incubator, put it under a microscope, focused, and said, Look here: La Crosse virus. I saw monkey cells, in a culture medium the color of cherry Kool-Aid, under attack by something so tiny it could only be discerned by the damage it did. People around the world—doctors, veterinarians—send him tissue samples, Calisher explained, asking him to grow a virus from the stuff and identify it. Okay. That sort of thing has been his life's work, especially with regard to hantaviruses in rodents. And then came this little excursion into bats.

We repaired to his office, now almost empty as he eased into retirement, except for a desk, two chairs, a computer, and some boxes. He tilted back in his chair, set his boots on the desk, and began to talk: arboviruses, the CDC, hantaviruses in rodents, La Crosse virus, mosquitoes, and a congenial group called the Rocky Mountain Virology Club. He ranged widely but, knowing my interest, circled back to a consequential chat he'd had with a colleague about six years earlier, soon after news broke that SARS, the new killer coronavirus, had been traced to a Chinese bat. The colleague was Kathryn V. Holmes, an expert on coronaviruses and their molecular structure, at the University of Colorado Health Sciences Center near Denver, just down the highway from Fort Collins. Charlie told me the story in his own vivid way, complete with dialogue:

"We oughta write a review paper about bats and their viruses," he said to Kay Holmes. "This bat coronavirus is really interesting."

She seemed intrigued but a little dubious. "What would we include?"

"Well, this and that, something else," Charlie said vaguely. The idea was still taking shape. "Maybe immunology."

"What do we know about immunology?"

Charlie: "I don't know shit about immunology. Let's ask Tony."

Tony Schountz, another professional friend, is an immunologist, then at the University of Northern Colorado, in Greeley, who does research on responses to hantaviruses in humans and mice. At that time Schountz, like Calisher, had never studied chiropterans. But he was a burly young guy, a former athlete, who had played college baseball as a catcher.

"Tony, what do you know about bats?"

Schountz thought Charlie meant baseball bats. "They're made of ash."

"Hello, Tony? I'm talkin' about *bats*." Wing-flapping gesture. As distinct from: Joe DiMaggio gesture.

"Oh. Uh, nothing."

"You ever read anything about the immunology of bats?"

"No."

"Have you ever *seen* any papers on the immunology of bats?"

"No."

Neither had Charlie—nothing beyond the level of finding antibodies that confirmed infection. Nobody seemed to have addressed the deeper question of how chiropteran immune systems respond. "So I said to Kay, 'Let's write a review paper,'" Charlie told me. "Tony said, 'Are you crazy? We don't *know* anything.'"

"Well, *she* doesn't know anything, *you* don't know anything, and *I* don't know anything. This is great. We don't have any biases."

"*Biases?*" said Schountz. "We don't have any *information!*"

"I said, 'Tony, that shouldn't hold us back.'"

Thus the workings of science. But Calisher and his two pals didn't plan to flaunt their ignorance. If we don't know anything in this or that area, he proposed, we'll just get somebody who *does*. They enlisted James E. Childs, an epidemiologist and rabies expert at the Yale School of Medicine (and an old friend of Charlie's from CDC days), and Hume Field, an Australian veterinary ecologist, who had played a key role in the identification of the bat reservoirs of both Nipah and Hendra. This five-member team, with their patchwork of expertise and their sublime lack of biases, then wrote a long, wide-ranging paper. Several journal editors voiced interest but wanted the manuscript cut; Charlie refused. It appeared finally, intact, in a more expansive journal, under the title "Bats: Important Reservoir Hosts of Emerging Viruses." It was a review, as Charlie had envisioned, meaning that the five authors made no claim of presenting original research; they simply summarized what had previously been done, gathered disparate results together (including unpublished data contributed by others), and sought to highlight some broader patterns. That much, it turned out, was a timely service. The paper offered a rich compendium of facts and ideas—and where facts were scarce, directive questions. Other disease scientists noticed. "All of a sudden," Charlie told me, "the phone's ringing off the hook." They met hundreds of requests for reprints, maybe thousands, sending their "Bats: Important Reservoir Hosts" to colleagues worldwide in the form of a PDF. Everybody wanted to know—everybody in that professional universe anyway—about these new viruses and their chiropteran hideouts. Yes, what *is* the deal with bats?

The paper made a handful of salient points, the first of which put the rest in perspective: Bats come in many, many forms. The order *Chiroptera* (the "hand-wing" creatures) encompasses 1,116 species, which amounts to 25 percent of all the recognized species of mammals. To say again: One in every four species of mammal is a bat. Such diversity might suggest that bats *don't* harbor more than their share of viruses; it could be, instead, that their viral burden is proportional to their share of all mammal diversity, and thus just *seems* surprisingly large. Maybe their virus-per-species ratio is no higher than ratios among other mammals.

Then again, maybe it *is* higher. Calisher and company explored some reasons why that might be so.

Besides being diverse, bats are very abundant and very social. Many kinds roost in huge aggregations that can include millions of individuals at close quarters. They are also a very old lineage, having evolved to roughly their present form about 50 million years ago. Their ancientness provides scope for a long history of associations between viruses and bats, and those intimate associations may have contributed to viral diversity. When a bat lineage split into two new species, their passenger viruses may have split with them, yielding more kinds of virus as well as more kinds of bats. And the abundance of bats, as they gather to roost or to hibernate, may help viruses to persist in such populations, despite acquired immunity in many older individuals. There's a concept in disease ecology called "critical community size" (CCS); a population size for hosts below which infectious pathogens tend to die out, because they simply don't have enough susceptible individuals to sustain them. Measles virus, for instance, seems to have a critical community size of roughly 250,000 humans; in an isolated human population smaller than that (on an island, for instance), the virus disappears after everyone has been exposed. Bats probably meet the CCS standard more consistently than most other mammals. Their communities are often huge and usually large, offering a steady supply of susceptible newborns to become infected and maintain the viral presence.

That scenario assumes a virus that infects each bat only briefly, leaving recovered individuals with lifelong immunity, as measles does in humans. An alternative scenario involves a virus capable of causing chronic, persistent infection, lasting months or even years within a single bat. If the infection can persist, then the long average lifespan of a bat becomes advantageous for the virus. Some of the smaller, insectivorous bats live twenty or twenty-five years. Such longevity, if the bat is infected and shedding virus, vastly increases the sum of opportunities over time for passing the virus to other bats.

Social intimacy helps too, and many kinds of bat seem to love crowding, at least when they hibernate or roost. Mexican free-tailed bats in Carlsbad Caverns, for instance, snuggle together at about three hundred individuals per square foot. Not even lab mice in an overloaded cage would tolerate that. If a virus can be passed by direct contact, bodily fluids, or tiny droplets sprayed through the air, crowding improves its chances. Under conditions like those in Carlsbad, Calisher's group noted, even rabies has been known to achieve airborne transmission.

Speaking of airborne: It's not insignificant that bats fly. An individual fruit bat may travel dozens of miles each night, searching for food, and hundreds of miles in a season as it moves among roosting sites. Some insectivorous bats migrate as much as eight hundred miles between their summer and winter roosts. Rodents don't make such journeys, and not many larger mammals do. Furthermore, bats move in three dimensions across the landscape, not just two; they fly high, they swoop low, they cruise in between, inhabiting a far greater volume of space than most animals. The breadth and the depth of their sheer presence are large. Does that increase the likelihood that they, or the viruses they carry, will come in contact with humans? Maybe.

Then there's bat immunology. Calisher's group could only touch judiciously on this topic, even with Tony Schountz as a coauthor, because little is known by anyone. Mainly they raised questions. Is it possible that the cold temperatures endured by hibernating bats suppress their immune responses, allowing viruses to persist in bat blood? Is it possible that antibodies, which would neutralize a virus, don't last as long in bats as in other mammals? What about the ancientness of the bat lineage? Did that lineage diverge from other mammals before the mammalian immune system had been well honed by evolution, reaching the level of effectiveness seen in rodents and primates? Do bats have a different "set point" for their immune responses, allowing a virus to replicate freely so long as it doesn't do the animal any harm?

Answering those questions, according to Calisher's group, would require new data derived from new work. And that work couldn't be done just with the sleek tools and methods of molecular genetics, comparing long sequences of nucleotide bases on the DNA or RNA molecule by way of computer software. They wrote:

Emphasis, sometimes complete emphasis, on nucleotide sequence characterization rather than virus characterization has led us down a primrose path at the expense of having real viruses with which to work.

The paper was a collaborative effort but that sentence sounds like Charlie Calisher. What it means is: *Hello, people? We've gotta grow these bugs the old-fashioned way, we've gotta look at them in the flesh, if we're gonna understand how they operate.* And if we don't, the paper added, "we are simply waiting for the next disastrous zoonotic virus outbreak to occur."

17

Charlie Calisher and his coauthors, besides touching on broad principles, discussed a handful of bat-related viruses in detail: Nipah, Hendra, rabies and its close relatives (the lyssaviruses), SARS coronavirus, and a couple of others. They mentioned Ebola and Marburg, though carefully omitting those two from the list of viruses for which bats had been proven to serve as reservoirs. "The natural reservoir hosts of these viruses have not yet been identified," they said about Marburg and Ebola—accurately, as of the time of publication. Their paper appeared in 2006. Fragments of Ebola RNA had recently been detected, by Eric Leroy and his colleagues, in some bats; antibodies against Ebola virus had been found in other bats. But that wasn't quite proof enough. Nobody had yet isolated any live filovirus from a bat, and the unsuccessful efforts to do so left Ebola and Marburg well hidden.

Then, in 2007, Marburg virus reappeared, this time among miners in Uganda. It was a small outbreak, affecting only four men, of whom one died, but it served as an opportunity to gain new insight on the virus, thanks in part to a quickly responsive multinational team. The four victims all worked at a site called Kitaka Cave, not far from Queen Elizabeth National Park, in the southwestern corner of Uganda. They dug galena, which is lead ore, plus a little bit of gold. The word "mine" caught the attention of some scientists within the CDC's Special Pathogens Branch, in Atlanta, because they already had reasons to suspect that Marburg's reservoir, whatever it was, might be associated with cavelike environments. Several of the previous Marburg outbreaks included patients whose cases histories involved visits to, or work in, caves or mines. So when the response team arrived at Kitaka Cave, in August 2007, they were ready to go underground.

This group included scientists from the CDC, the National Institute for Communicable Diseases in South Africa, and the WHO in Geneva. The CDC sent Pierre Rollin and Jonathan Towner, whom we've met before, as well as Brian Amman and Serena Carroll. Bob Swanepoel and Alan Kemp of the NICD flew up from Johannesburg; Pierre Formenty arrived from the WHO. All of them possessed extensive experience with Ebola and Marburg, gained variously through outbreak responses, lab research, and field studies. Amman was a mammalogist with a special affinity for bats. During a conversation at the CDC, he described to me what it was like to go into Kitaka Cave.

The cave served as roosting site for about a hundred thousand individuals of the Egyptian fruit bat (*Rousettus aegyptiacus*), a prime suspect as reservoir for Marburg. The team members, wearing Tyvek suits, rubber boots, goggles, respirators, gloves, and helmets, were shown to the shaft by miners, who as usual were clad only in shorts, T-shirts, and sandals. Guano covered the ground. The miners clapped their hands to scatter low-hanging bats as they went. The bats, panicked, came streaming out. These were sizable animals, each with a two-foot wingspan, not quite so large and hefty as the flying foxes of Asia but still daunting, especially with thousands swooshing at you in a narrow tunnel. Before he knew it, Amman had been conked in the face by a bat and suffered a cut over one eyebrow. Towner got hit too, Amman said. Fruit bats have long, sharp thumbnails. Later, because of the cut, Amman would get a postexposure shot against rabies, though Marburg was a more immediate concern. "Yeah," he thought, "this could be a really good place for transmission."

The cave had several shafts, Amman explained. The main shaft was about eight feet high. Because of all the mining activity along there, many of the bats had shifted their roosting preference "and went over to what we called the cobra shaft." That was a smaller shaft, branching off, which—

I interrupted him. "'Cobra' because there were *cobras*?"

"Yeah, there was a black forest cobra in there," he said.

Or maybe a couple. It was good dark habitat for snakes, with water and plenty of bats to eat. Anyway, the miners showed Amman and Towner into the cave, past another narrow shaft that led to a place called the Hole, a pit about ten feet deep accessed by shinnying down a pole, from the bottom of which came much of the ore. The two Americans were looking for the Hole but, following their guides, inadvertently passed that shaft by, continuing about two hundred meters along the main shaft to a chamber containing a body of brown, tepid water. Then the local fellows cleared out, leaving Towner and Amman to do a bit of exploring on their own. They dropped down beside the brown lake and found that the chamber branched into three shafts, each of which seemed blocked by standing water. Peering into those shafts, they could see many more bats. The humidity was high and the temperature maybe ten or fifteen degrees hotter than outside. Their goggles fogged up. Their respirators became soggy and wouldn't pass much oxygen. They were panting and sweating, zipped into their Tyvek suits, which felt like wearing a trash bag, and by now they were becoming "a little loopy," Amman recalled. One lakeside shaft seemed to curve back around, possibly connecting with the cobra shaft. They didn't know how deep the water might be, and the airspace above it was limited. Should they proceed? No, they decided, the increased risk wasn't worth the potential benefit. Formenty, their WHO colleague, eventually found them down there and said, Hey, guys, the Hole is back this way. They crawled out and retraced their path, "but by that time we were spent," Amman said. "We had to get out and cool off." It was only their first underground excursion at Kitaka. They would make several.

On a later day, the team investigated a grim, remote chamber they dubbed the Cage. It was where one of the four infected miners had been working just before he got sick. This time Amman, Formenty, and Alan Kemp of the NICD went to the far recesses of the cave. The Cage itself could only be entered by crawling through a low gap at the base of a wall—like sliding under a garage door that hadn't quite closed. Brian Amman is a large man, six foot three and 220 pounds, and for him the gap was a tight squeeze; his helmet got stuck and he had to pull it through separately. "You come out into this sort of blind room," he said, "and the first thing you see is just hundreds of these dead bats."

They were Egyptian fruit bats, the creature of interest, left in various stages of mummification and rot. Piles of dead and liquescent bats seemed a bad sign, potentially invalidating the hypothesis that Egyptian fruit bats might be a reservoir host of Marburg. If these bats had died in masses from the virus, then they couldn't also be its reservoir. Then again, they might have succumbed to earlier efforts by the locals to exterminate them with fire and smoke. Their cause of death was indeterminable without more evidence, and that's partly why the team was there. If these bats *had* died of Marburg, suspicion would shift elsewhere—to another bat, or maybe a rodent, or a tick, or a spider? Those other suspects might have to be investigated. Ticks, for instance: There were plenty of them in crevices near the bat roosts, waiting for a chance to drink some blood. Meanwhile, when Amman and Kemp stood up in the Cage, they realized that not every bat in there was dead. The room was aswirl with live ones, circling around their heads.

The two men went to work, collecting. They stuffed dead bats into bags. They caught a few live bats and bagged them too. Then, back down on their bellies, they squooched out through the low gap. "It was really unnerving," Amman told me. "I'd probably never do it again." One little accident, he said, a big rock rolls in the way, and that's it. You're trapped.

Wait a minute, lemme get this straight: You're in a cave in Uganda, surrounded by Marburg and rabies and black forest cobras, wading through a slurry of dead bats, getting hit in the face by live ones like Tippi Hedren in *The Birds*, and the walls are alive with thirsty ticks, and you can hardly breathe, and you can hardly see, and . . . you've got time to be *claustrophobic?*

"Uganda is not famous for its mine rescue teams," he said.

By the end of this field trip, the scientists had collected about eight hundred bats for dissection and sampling, half of those belonging to *Rousettus aegyptiacus*. The CDC team, including Towner and Amman, returned to Kitaka Cave seven months later, in April 2008, catching and sampling two hundred more individuals of *R. aegyptiacus* to see if Marburg persisted in the population. If so, that would strongly suggest that this species was in fact a reservoir. During the second trip, they also marked and released more than a thousand bats, hoping that from later recaptures they could deduce the overall size of the population. Knowing the population size, as well as the prevalence of infection among their sampled bats, would indicate how many infected bats might be roosting in Kitaka at any one time. Towner and Amman used beaded collars (which seemed less discomfiting to the bats than the usual method of marking, leg bands), each collar coded with a number. The two scientists took some heat for this mark-recapture study; skeptical colleagues argued that it was wasted effort, given the vast size of the bat population and the odds against recapture. But, in Amman's words, "we kind of stuck to our guns," and they eventually released 1,329 tagged bats.

Less speculative, less controversial, were the samples of blood and tissue from dissected bats. Those went back to Atlanta, where Towner took part in the laboratory efforts to find traces of Marburg virus. One year later came a paper, authored by Towner, Amman, Rollin, and their WHO and NICD colleagues, announcing some important results. All the cave crawling, bat sampling, and lab work had yielded a dramatic breakthrough in the understanding of filoviruses, meaning both Marburg and Ebola. Not only did the team detect antibodies against Marburg (in thirteen of the roughly six hundred fruit bats sampled) and fragments of Marburg RNA (in thirty one of the bats), but they also did something more difficult and compelling. Antibodies and RNA fragments, though significant, were just the same sorts of secondary evidence that had provisionally linked the Ebola virus to bats. This team had gone a step farther: They'd found live virus.

Working in one of the CDC's BSL-4 units, Towner and his co-workers had isolated viable, replicating Marburg virus from five different bats. Furthermore, the five strains of virus were genetically diverse, suggesting an extended history of viral presence and evolution within Egyptian fruit bats. Those data, plus the fragmentary RNA, constituted strong evidence that the Egyptian fruit bat is a reservoir—if not *the* reservoir—of Marburg virus. Based on the isolation work, it's definitely there in the bats. Based on the RNA fragments, it seems to infect about 5 percent of the bat population at a given time. Putting those numbers together with the overall population estimate of a hundred thousand bats at Kitaka, the team could say that about five thousand Marburg-infected bats flew out of the cave every night.

An interesting thought: five thousand infected bats passing overhead. Where were they going? How far to the fruiting trees? Whose livestock or little gardens got shat upon as they went? Jon Epstein had warned me, during our trapping adventures in Bangladesh, about gaping at overhead bats that were presumed reservoir hosts of a lethal virus: "Keep your mouth closed when you look up." Salubrious advice also for the farmers and villagers roundabout Kitaka cave. And that aggregation, Towner and his coauthors added, "is only one of many such cave populations throughout Africa."

Where else might Marburg virus be traveling on the wings of these bats? An answer to that arrived the following summer.

18

Astrid Joosten was a 41-year-old Dutch woman who, in June 2008, went to Uganda with her husband on an adventure vacation. It wasn't their first but it would be her last.

At home in North Brabant, a southern province of the Netherlands, Joosten worked as a business analyst for an electrical company. Both she and her spouse, a financial manager, enjoyed escaping on annual getaways to experience the landscapes and cultures of other countries, especially in Africa. In 2002 they had flown to Johannesburg and, stepping off the airplane, felt love at first sight. On later trips they visited Mozambique, Zambia, and Mali. The journey in 2008, booked through an adventure-travel outfitter, would allow them to see mountain gorillas in the southwestern highlands of the country as well as some other wildlife and cultures. They worked their way south toward Bwindi Impenetrable Forest, where the Ugandan gorillas reside. On one intervening day, the operators offered a side trip, an option, to a place called the Maramagambo Forest, where the chief attraction was a peculiar site that everyone knew as Python Cave. African rock pythons lived there, languid and content, grown large and fat on a diet of bats.

Joosten's husband, later her widower, was a fair-skinned man named Jaap Taal, a calm fellow with a shaved head and dark, roundish glasses. Most of the other travellers didn't fancy this offering, Jaap Taal told me, over a cup of coffee at a café in southwestern Montana. Never mind, for the moment, why he turned up there. Python Cave had been an add-on, he explained, price not included in their Uganda package. "But Astrid and I always said, maybe you come here only once in your life, and you have to do everything you can." They rode to Maramagambo Forest and then walked a mile or so, gradually ascending, to a small pond. Nearby, half-concealed by moss and other greenery, like a crocodile's eye barely surfaced, was a low dark opening. Joosten and Taal, with their guide and one other client, climbed down into the cave.

The footing was bad: rocky, uneven, slick with bat guano. The smell was bad too: fruity and sour. Think of a dreary barroom, closed and empty, with beer on the floor at 3 a.m. The cave seemed to have been carved by a creek, or at least to have channeled its waters, and part of the overhead rock had collapsed, leaving a floor of boulders and coarse rubble, a moonscape, coated with guano like a heavy layer of vanilla icing. The ceiling was thick with bats, big ones, many thousands of them, agitated and chittering at the presence of human intruders, shifting position, some dropping free to fly and then settling again. Astrid and Jaap kept their heads low and watched their steps, trying not to slip, ready to put a hand down if needed. "I think that's how Astrid got infected," he told me. "I think she put her hand on a piece of rock, which contained droppings of a bat, which were infected. And so she had it on her hand." Maybe she touched her face an hour later, or put a piece of candy in her mouth, or something such, "and that's how I think the infection got in her."

Python Cave, in Maramagambo Forest, is just thirty miles west of Kitaka Cave. It too harbors Egyptian fruit bats. Thirty miles isn't far and individuals from the Kitaka aggregation are quite capable—as the CDC's mark-recapture study would later prove—of finding their way to roost at Python.

No one had warned Joosten and Taal about the potential hazards of an African bat cave. They knew nothing of Marburg virus, though they had heard of Ebola. They only stayed in the cave about ten minutes. They saw a python, large and torpid. Then they left, continued their Uganda vacation, visited the mountain gorillas, did a boat trip, and flew back to Amsterdam. Thirteen days after the cave visit, home in North Brabant, Astrid Joosten fell sick.

At first it seemed no worse than flu. Then her temperature went higher and higher. After a few days, she began suffering organ failure. Her doctors, knowing her case history, with recent time in Africa, suspected Lassa virus or maybe Marburg. Marburg, said Jaap, what's that? Astrid's brother looked it up on Wikipedia and told him: Marburg, it kills, could be bad trouble. The doctors moved her to a hospital in Leiden, where she could get better care and be isolated from other patients. There she developed a rash and conjunctivitis; she hemorrhaged. She was put into an induced coma, a move dictated by the need to dose her more aggressively with antiviral medicine. Before she lost consciousness, though not long before, Jaap went back into the isolation room, kissed her, and told her, "Well, we'll see you in a few days." Blood samples, sent to a lab in Hamburg, confirmed the diagnosis: Marburg. She worsened. As her organs shut down, she lacked for oxygen to the brain, she suffered cerebral edema, and before long Astrid Joosten was declared brain dead. "They kept her alive for a few more hours, until the family arrived," Jaap told me. "Then they pulled the plug out and she died within a few minutes."

The doctors, appalled by his recklessness in kissing her goodbye, prepared an isolation room for Jaap himself, but it was never needed. "There's so much they don't know about Marburg and those other viral infections," he said to me. Then, still a venturesome traveler, he went off on a snow tour of Yellowstone National Park.

19

News of the Joosten case reverberated at the CDC. She was the first person known to have left Africa with an active filovirus infection and died. Soon afterward, in August 2008, another team was dispatched to Uganda, this time including the veterinary microbiologist Tom Ksiazek, a veteran of field responses against zoonotic outbreaks, as well as Towner and Amman. Bob Swanepoel and Alan Kemp were again mustered from South Africa. "We got the call, 'Go investigate,'" Amman told me. Their mission now was to sample bats at Python Cave, where this Dutch woman (unnamed in the epidemiological traffic) had become infected. Her death, her case history, implied a change in the potential scope of the situation. That local Ugandans were dying of the infection was a severe and sufficient concern—sufficient to bring a response team in haste from Atlanta and Johannesburg. But if tourists too were involved, were tripping in and out of some lovely python-infested Marburg repository, in Tevas and hiking boots, blithe, unprotected, and then boarding their return flights to other continents, the place was not just a peril for Ugandan miners and their families. It was also an international threat.

The team converged at Entebbe and drove southwest. They walked the same trail that Joosten and Taal had walked, to the same crocodile-eye opening amid the forest vegetation. Then, unlike any tourists, they donned their Tyvek pajamas, their rubber boots, their respirators, and their goggles. This time, with cobras in mind, they added snake chaps. Then they went in. Bats were everywhere overhead; guano was everywhere underfoot. In fact, the rain of guano seemed to come so continuously, Amman told me, that if you left something on the floor it would be covered within days. The pythons were indolent and shy, as well-fed snakes tend to be. One of them, by Amman's estimate, stretched about twenty feet long. The black forest cobras (yes, more of them here too) kept to the deeper recesses, away from heavy traffic. Towner was gazing at a python when Amman noticed something glittery on the floor.

At first glance it looked like a bleached vertebra, lying in the excremental glop. Amman picked the thing up.

It wasn't a vertebra. It was a string of aluminum beads with a number attached. More specifically, it was one of the beaded collars that he and Towner had placed on captured bats at Kitaka Cave, the *other* Marburg cave, three months earlier and thirty miles away. The code tag spoke one simple fact: Here was collar K-31, from the thirty-first animal they had released. "And of course, I just lost my mind," Amman told me. "I was, 'Yeah!' and jumping around. Jon and I were so excited." Amman's insane jubilance was in fact just the sane, giddy thrill that a scientist feels when two small bits of hard-won data click together and yield an epiphany. Towner got it and shared it. Picture two guys in a dark stone room, wearing headlamps, high-fiving in nitrile gloves.

Retrieving the collar at Python Cave vindicated, in a stroke, their mark-recapture study. "It confirmed my suspicions that these bats are moving," Amman said—and moving not only through the forest but from one roosting site to another. Travel of individual bats (such as K-31) between far-flung roosts (such as Kitaka and Python) implied circumstances whereby Marburg virus might ultimately be transmitted all across Africa, from one bat encampment to another. It suggested opportunities for infecting or re-infecting bat populations in sequence, like a string of blinking Christmas lights. It voided the comforting assumption that this virus is strictly localized. And it highlighted the complementary question: Why don't outbreaks of Marburg virus disease happen more often?

Marburg is only one disease to which that question applies. Why not more SARS? Why not more Nipah? Why not more Ebola? If bats are so abundant and diverse and mobile, and zoonotic viruses so common within them, why don't those viruses spill into humans and take hold more frequently? Is there some mystical umbrella that protects us? Or is it fool's luck?

20

Although I have lumped Ebola virus provisionally, along with Marburg and SARS coronavirus and Nipah and others, among viruses for which bats serve as reservoirs, I want to re-emphasize: That inclusion is tentative. It's a hypothesis awaiting assessment against further evidence. No one, as of this writing, has isolated any live ebolavirus from a bat—and virus isolation is still the gold standard for identifying a reservoir.

That may happen soon; people are trying. Meanwhile the Ebola-in-bats hypothesis seems stronger since Jonathan Towner's team achieved their isolations of Marburg virus, so closely related, also from bats. And it has been strengthened further, at least a little, by another bit of data added to the ebolavirus dossier about the same time. This bit came in the form of a story about a little girl.

Eric Leroy, after having chased the secrets of Ebola for more than a decade, led the team that reconstructed this girl's story. Their new evidence derived not from molecular virology but from old-fashioned epidemiological detective work—interviewing survivors, tracing contacts, discerning patterns. The context was an outbreak of Ebola virus that occurred in and around a village called Luebo, along the Lulua River, in a southern province of the Democratic Republic of the Congo. Between late May and November 2007, at least 264 people sickened with what seemed to be or (in some confirmed cases) definitely was Ebola virus. Most of them died. The lethality was 70 percent. Leroy and his colleagues arrived in October, as part of an international WHO response team in cooperation with the DRC's Ministry of Health. Leroy's study focused on the network of transmissions, which all seemed traceable to a certain 55-year-old woman. She became known, in their report, as patient A. She wasn't necessarily the first human to get infected; she was merely the first identified. This woman, elderly by Congo village standards, had died after suffering high fever, vomiting, diarrhea, and hemorrhages. Eleven of her close contacts, mainly family, who had helped care for her, sickened and died too. The outbreak spread onward from there.

Leroy and his group wondered how the woman herself had gotten infected. No one in her village had shown symptoms before she did. So the investigators broadened their search to surrounding villages, of which there were quite a few, both along the river and in the forest nearby. From their interviews and their legwork, they learned that the villages were interconnected by footpaths, and that on Mondays the heavy traffic led to one particular village,

Mombo Mounene 2, the site of a big weekly market. They also learned about an annual aggregation of migrating bats.

The bats generally arrived in April and May, stopping over amid a longer journey, finding roost sites and wild fruit trees on two islands in the river. In an average year, there might be thousands or tens of thousands of animals, according to what Leroy's group heard. In 2007, the migration was especially large. From their island roosts, the bats ranged the area. Sometimes they fed at a palm oil plantation along the river's north bank; the plantation was a leftover from colonial times, now abandoned and gone derelict, but still offering palm fruits in April on its remaining trees. Many or most of the animals were hammer-headed bats and Franquet's epauletted fruit bats (*Epomops franqueti*), two of the three in which Leroy had earlier found Ebola antibodies. While roosting, the bats dangled thickly on tree branches. Local people, hungry for protein or a little extra cash, hunted them with guns. Hammer-headed bats, big and meaty, were especially prized. A single shotgun blast could bring down several dozen bats. Many of those animals ended up, freshly killed, raw and bloody, in the weekly market at Mombo Mounene 2, from which buyers carried them home for dinner.

One man who regularly walked from his own village to the market, and often bought bats, seems to have suffered a mild case of Ebola. The investigators eventually labeled him patient C. He wasn't a bat hunter himself; he was a retail consumer. During late May or early June, according to patient C's own recollection, he weathered some minor symptoms, mainly fever and headache. He recovered, but that wasn't the end of it. "Patient C was the father of a 4-year-old girl (patient B)," Leroy and his team later reported, "who suddenly fell ill on 12 June and died on 16 June 2007, having had vomiting, diarrhoea, and high fever." The little girl didn't hemorrhage, and she was never tested for Ebola, but it's the most plausible diagnosis.

How had she contracted it? Possibly she had shared in eating a fruit bat that carried the virus. What are the odds faced by bat-eaters? Hard to say; hard even to guess. If the hammer-headed bat *is* an Ebola reservoir, what's the prevalence of the virus within a given population? That's another unknown. Towner's group found 5 percent prevalence of Marburg in Egyptian fruit bats, meaning that one animal in twenty could be infected. Assuming a roughly similar prevalence in the hammer-headed bat, the little girl's family had been unlucky as well as hungry. They might have eaten nineteen other bats and gotten no exposure. Then again, if a bat meal was shared, why didn't the girl's mother and other family members get sick? Possibly her father, infected or besmeared after purchasing bats in the market, had carried the girl (common practice with small children thereabouts) along the footpath back to their village. The father, patient C, seems to have passed the virus to nobody else.

But his little daughter did pass it along. Her dead body was washed for burial, in accord with local traditions, by a close friend of the family. That friend was the 55-year-old woman who became patient A.

"Thus, virus transmission may have occurred when patient A prepared the corpse for burial ceremony," Leroy's group wrote. "When interviewed, the two other preparers, the girl's mother and grandmother, reported they did not have direct contact with the corpse and they did not develop any clinical sign of infection in the four following weeks." Their role in the funerary washing was apparently observational. They didn't touch the dead body of their daughter and granddaughter. But patient A did, performing faithfully the service of a close family friend, after which she went back to her life—what was left of it. She resumed her social interactions, for a time, and eventually 184 other people caught Ebola and died.

Leroy's team reconstructed this story and then, keen to extract meaning, asked themselves several questions. Why had the father infected his daughter but no one else? Maybe because he had a mild case, with a low level of virus in his body and not much leaking out. But if his case was mild, why was his daughter's so severe, killing her within four days? Maybe because, as a small child racked with vomiting and diarrhea, she had died of untreated dehydration.

Why was there only one bat-to-human spillover event? Why was patient C unique, as the sole case linked directly to the reservoir? Well, maybe he wasn't. He was just the only one that came to notice. "In fact, it is highly likely that several other persons were infected by bats," Leroy's group wrote, "but the circumstances required for subsequent human-to-human transmission were not present." They were alluding to dead-end infections. A person sickens, suffers solitarily or with carefully distanced succor from wary family or friends (food and water left at the door of a hut), and dies. Is buried unceremoniously. Eric Leroy didn't know how many unfortunate people in the Luebo area may have eaten a bat, touched a bat, become infected with Ebola virus, succumbed to it, and been dropped into a hole, having infected no one else. Amid the horrific confusion of the outbreak, in those remote villages, the number of such dead-end cases might have been sizable.

This brought Leroy's team to the pivotal question. If the circumstances required for human-to-human transmission hadn't been met, what *were* those circumstances? Why hadn't the Luebo outbreak gone really big? Why hadn't the tinder ignited the logs? It had started in May, after all, and the WHO didn't get there until October.

Whatever circumstances had combined to constrain the Luebo outbreak to a "mere" 264 cases and 186 fatalities, those circumstances would not be present during the West African outbreak of 2014.

21

This brings me back to that fruitless stakeout near Moba Bai, in the Republic of the Congo, in October 2006. It brings me back to the shared fates of people and apes.

After our days of searching the bai complex for gorillas,

and finding virtually none. Billy Karesh and I and the expert gorilla tracker Prosper Balo, along with other members of the team, traveled three hours back down the Mambili River by pirogue. We carried no samples of frozen gorilla blood, but I was nevertheless glad to have had the chance to come looking. From the lower Mambili we turned upstream on one of its branches, motored to a landing, and then drove a dirt road to the town of Mbomo, central to the area where Ebola virus had killed 128 people during the 2002–2003 outbreak.

Mbomo is where the medical anthropologist Barry Hewlett, arriving just after four teachers were hacked to death, had encountered murderous suspicions between one resident and another that the Ebola deaths resulted from sorcery. We stopped at a little hospital there, a U-shaped arrangement of low concrete structures surrounding a dirt courtyard, like a barebones motel. Each of the rooms, tiny and cell-like, gave directly onto the courtyard through a louvered door. As we stood in the heat, Alain Ondzie told me that Mbomo's presiding physician, Dr. Catherine Atsangandako, had famously locked an Ebola patient into one of those cells just a year earlier, supplying him with food and water through the slats. The man was a hunter, presumably infected by handling one form or another of wild meat. He had died behind his louvered door, a lonely end, but the doctor's draconian quarantine was generally credited with having prevented a wider outbreak.

Dr. Catherine herself was out of town today. The only evidence of her firm hand was a sign, painted in stark red letters:

> **ATTENTION EBOLA**
>
> **NE TOUCHONS JAMAIS**
>
> **NE MANIPULONS JAMAIS**
>
> **LES ANIMAUX TROUVES**
>
> **MORTS EN FORET**

Don't touch dead animals in the forest.

Mbomo had another small distinction: It was Prosper Balo's hometown. We visited his house, walking to it along a narrow byway and then a grassy path, and found its dirt courtyard neatly swept, with wooden chairs set out for us under a palm. We met his wife, Estelle, and some of his many children. His mother offered us palm whiskey. The children jostled for their father's attention; other relatives gathered to meet the strange visitors; we took group photos. Amid this cheery socializing, in response to a few gentle queries, we learned some details about how Ebola had affected Estelle and her family during that grim period in 2003, when Prosper had been away.

We learned that her sister, two brothers, and a child had all died in the outbreak, and that Estelle herself was shunned by townspeople because of her association with those fatalities. No one would sell food to her. No one would touch her money. Whether it was infection they feared, or dark magic, is uncertain. She had to hide in the forest. She would have died herself, Prosper said, if he hadn't taught her the precautions he'd learned from Dr. Leroy and the other scientists, around that time, while helping them in their search for infected animals: Sterilize everything with bleach, wash your hands, and don't touch corpses. But now the bad days were past and, with Prosper's arm around her, Estelle was a smiling, healthy young woman.

Prosper remembered the outbreak in his own way, mourning Estelle's losses and some of a different sort. He showed us a treasured book, like a family bible—except it was a botanical field guide—on the endpapers of which he had written a list of names: Apollo, Cassandra, Afrodita, Ulises, Orfeo, and almost twenty others. They were gorillas, an entire group that he had known well, that he had tracked daily and observed lovingly at the Lossi Gorilla Sanctuary, when he worked there. Cassandra was his favorite, Prosper said. Apollo was the silverback. "*Sont tous disparus en deux-mille trois,*" he said. All of them, gone in the 2003 outbreak. In fact, though, they hadn't entirely *disparus*: He and other trackers had followed the group's final trail and found six gorilla carcasses along the way. He didn't say which six.

Cassandra, dead with others in a fly-blown pile? It was very hard, he said. He had lost his gorilla family, and also members of his human family.

For a long time Prosper stood holding the book, opened for us to see those names. He comprehended emotionally what the scientists who study Ebola virus and other zoonoses know from their careful observations, their models, their data. People and gorillas, chimpanzees and bats, rodents and monkeys and viruses: We're all in this together.

EPILOGUE

On August 8, 2014, the World Health Organization declared the Ebola virus disease outbreak in West Africa to be a Public Health Emergency of International Concern (PHEIC). This seemed a truism to some attentive observers, and yet it was shocking to see the WHO make such a declaration in cold print. The announcement reflected the realities of a situation that, for a variety of reasons, had gone beyond the boundaries of previous experience with Ebola virus, so that even the experts were now on unfamiliar terrain.

Guinea had been the first country affected, then Liberia and Sierra Leone, and then by dire happenstance the virus travelled via airplane to Lagos, Nigeria, a raucous city of twenty-one million people. Elsewhere the numbers of cases and fatalities continued to rise quickly. Liberia was especially hard hit. Sierra Leone called out a battalion of soldiers to ensure that cases would remain isolated at treatment centers. In Guinea, amid rumors that health workers had deliberately spread the virus, riots broke out and young men with knives and guns threatened to attack a hospital. On the morning of September 2, Dr. Thomas R. Frieden, director of the CDC in Atlanta, told a CNN interviewer that the outbreak was "spiraling out of control." This seemed as unnerving as the WHO's PHEIC announcement; Frieden is no alarmist. Three days later, United Nations Secretary-General Ban Ki-moon issued an "international rescue call" for a "massive surge in assistance" from the global community, warning that the disease "is spreading far faster than the response." By that time the total number of probable, confirmed, and suspected cases had reached 3,988, including 2,112 deaths—more cases, and more fatalities, than in all previously known outbreaks of Ebola combined. People around the world were worried and sympathetic and bewildered. People in Liberia and its neighboring countries were scared and angry and aggrieved. Many of them, too, were sick and dying.

That's where things stand as I write this, wondering which way the trends will have tipped by the time you read it. No one can predict just how much more awful the West Africa outbreak will become; there are too many variables, some of which are impossible to calculate, others calculable but (as Thomas Frieden suggested) almost beyond control. We don't even know whether the past is a reliable guide to the future—that is, to what degree history and science can illuminate the Ebola events of 2014. But I've tried to offer a bit of both in this little book, science and history, on the chance that they might provide useful context for what you're seeing, hearing, and reading in the news reports, and perhaps even make you slightly better equipped to act as global citizens in the face of what has become a global challenge. At very least, a bit of history and science can put in relief how present events might resemble, or differ from, those in the past, and why.

The outbreak seems to have begun as early as December 2013, in the Guéckédou prefecture of southern Guinea, not far from the borders of both Liberia and Sierra Leone. A two-year-old boy in a village called Meliandou began showing symptoms—fever, black stool, vomiting—and died four days later, on December 6. His mother hemorrhaged fatally the following week. Then his three-year-old sister sickened on Christmas, with symptoms similar to the boy's, and died quickly too. Their grandmother, again after fever and vomiting and diarrhea, died on January 1, 2014. From there the outbreak spread, evidently by way of family care-givers, foot travel, and contacts that may have occurred at the grandmother's funeral. It reached other villages, as well as hospitals in two nearby towns. A doctor in the town of Macenta, after attending one patient, took sick himself, with symptoms that included vomiting, bleeding, and hiccups, and soon died. The doctor's funeral brought the virus to yet another town.

All this happened without being noticed by national or international health authorities. Then, on March 10, 2014, health officials from the region alerted the Guinean Ministry of Health about the alarming clusters of illness and death. Guinean doctors and scientists from the capital, Conakry, now became involved; Médecins sans Frontières (Doctors without Borders) also sent a team, which began caring for patients and sending blood samples up to BSL-4 laboratories in France and Germany for analysis. A team drawn from all these professionals hurriedly compiled a scientific report, with Sylvain Baize of the Pasteur Institute in Lyon as first author, which was published online by *The New England Journal of Medicine* in April, charting the chain of infections and making one other signal point. The bug they had found was not Taï Forest virus, as might have been expected on grounds of geography (given that Côte d'Ivoire shares a border with Guinea). No, this outbreak was caused by a different ebolavirus, a variant of Ebola virus itself, as known from Gabon and the two Congo countries, roughly two thousand miles eastward. The Baize study also noted that the three kinds of fruit bat implicated by Eric Leroy's group as suspected Ebola virus reservoirs, including the hammer-headed bat, are present in parts of West Africa.

The hammer-headed bat, in fact, has a distribution that includes southern Guinea. Did the two-year-old boy in Meliandou, the apparent first case, contract his infection from a hammer-headed bat? It's possible but it isn't known.

Another important study of the genetics of the virus appeared in late August, in *Science* Express (a streamlined publication of the journal *Science*), under the authorship of Stephen K. Gire, of Harvard, and a long list of coauthors. Five of those coauthors, having worked amid the outbreak, had died of Ebola virus disease by the time this study was published, giving it a certain extra gravitas. Based on their sequencing of the genomes of virus samples from 78 patients in Sierra Leone, Gire and his colleagues reported three notable results. First, the virus was mutating prolifically and accumulating a fair degree of genetic variation as it replicated within each human case and passed from one human to another.

(So it was changing, evolving, as time progressed; whether it was adapting to humans is a separate but related question.) Second, the 78 samples were sufficiently similar to suggest that they had all descended from a single recent ancestor, implying just one spillover from the reservoir host. Third, comparative analysis of the samples showed this West African variant of Ebola virus to be distinct from Ebola virus as lately seen in the Democratic Republic of the Congo by about ten years worth of mutational differences. It had evidently evolved independently in its reservoir host for about a decade since becoming isolated from the Central African lineage.

That last finding, just ten years worth of localized mutation for the virus within its West African reservoir, seemed to suggest that Peter Walsh might be right. Is it possible that Ebola virus is still spreading like a wave through the bat populations of Central and West Africa—reaching new locations, diverging genetically, and presenting new dangers to people in those places? Maybe.

While these scientific studies progressed, so did the spread of the outbreak and the anguished but unsuccessful efforts to contain it. Those efforts were hampered by a number of factors: the weakness of governance in Liberia, Guinea, and Sierra Leone after decades of coups, juntas, and civil wars; the bitterness and suspicion among their peoples as legacy of those conflicts; the inadequacy of health-care infrastructure and basic health-care services in the three countries, as reflected in extremely low annual per capita expenditures on health; the immediate shortage of money and outbreak-response supplies necessary for stopping Ebola, such as examination gloves, masks, gowns, rubber boots, bleach, and plastic buckets in which to put bleach solution so that hands could be washed; the shortage of treatment centers and beds within them; the porosity of the national borders between Guinea, Liberia, and Sierra Leone; the reluctance of people in affected villages and towns to see their loved ones confined to isolation facilities, within which treatment was often marginal and case fatality rate was running above 50 percent; the reluctance of people to suspend their traditional burial practices, which often involved washing or otherwise touching the body; the relatively short distances between rural areas where the outbreak started and the capital cities of the three countries, allowing people to travel from affected areas to Monrovia, Freetown, and Conakry by such relatively inexpensive modes of transport as shared taxi and bus; and the shortage of timely international aid. Notwithstanding heroic efforts by many Liberian, Guinean, and Sierra Leonean doctors, nurses, and other health-care workers, and by the courageous foreign responders from Médecins sans Frontières, Samaritan's Purse, the CDC, the WHO, and other organizations, there just wasn't enough material support and expertise on the scene, not yet, to contain this unusually difficult outbreak.

And then the virus rode an airplane to Nigeria. Some observers, including myself, had noted in the past that Ebola virus doesn't ride airplanes well, because it tends to debilitate its victims so quickly, and because it typically affects poor people in remote villages amid forest areas, who can't afford air travel anyway; but we were wrong if we seemed to imply that it can't ride airplanes *at all*. On July 20, a Liberian-American man named Patrick Sawyer arrived in Lagos after a visit to Liberia, during which he had reportedly cared for his Ebola-stricken sister. He sickened during the last leg of the flight and was admitted to a hospital in Lagos, where he died on July 25. "Within days of his case being diagnosed, authorities in Nigeria were following 59 people who had had contact with the man," according to an account by Helen Branswell, a medical reporter for The Canadian Press, Canada's news agency. "The case count started to grow. A doctor tested positive. Then a nurse, who died." Although public health workers in Lagos traced Sawyer's contacts assiduously, one of them escaped the net and, defying a quarantine order, flew south to Port Harcourt, an oil-refining city of almost two million people on Nigeria's southern coast. There the escapee infected a doctor, who infected others, igniting a worrisome new hotspot of the outbreak.

The virus has now also reached Senegal, another West African neighbor, by way of a student from Guinea. It may well get to other locations, other countries, transported within infected people by bus, foot, or bicycle across open borders, or by air travel internationally or intercontinentally. So we have reached a point, as of September 2014, where we've got to stop calling this an outbreak and begin calling it an epidemic. Stephen K. Gire and his colleagues took that step in their genomic study of the virus.

The word "pandemic," so fearfully resonant, is still inappropriate and—with anything but the worst of luck—will continue to be inappropriate. Ebola as we know it is just not the right sort of virus to spread around the world, as influenza regularly does, and as another respiratory bug such as SARS coronavirus might, causing tens of thousands of deaths, not just in poor countries but also in wealthy ones with all the advantages of strong governance and rigorous health care. This is still a slow-moving virus, compared to many. What we should recognize, what we should remember, is that the events in West Africa (so far) tell us not just about the ugly facts of Ebola's transmissibility and lethality; they tell us also about the ugly facts of poverty, inadequate health care, political dysfunction, and desperation in three West African countries, and of neglectful disregard of those circumstances over time by the international community.

That said, though, I've got to mention one other dangerous factor we need to consider: evolution. As I mentioned above, the Gire study found a "rapid accumulation" of genetic variation in the virus from those 78 patients. The observed rate, in fact, was twice as high as the normal rate of mutation in Ebola virus between outbreaks. Furthermore, some or many of those mutations have been consequential ones ("nonsynonymous," in genetic lingo), changing the identity of an amino acid for which the RNA codes. Change like that can lead to functional changes. The high rate of nonsynonymous mutations, according to Gire and his colleagues, "suggests that continued progression of this epidemic could afford an opportunity for viral adaptation, underscoring the need for rapid containment." In plain language: The higher the case count goes, the greater the likelihood that Ebola virus as we know it might evolve into something better adapted to pass from human to human, something that presently exists only in our nightmares.

The other piece of grim but illuminating news, as I write this, is that *another* outbreak of Ebola virus disease has begun, far to the east, in a remote and forested northern province of the Democratic Republic of the Congo. The Congo outbreak involves a different variant of Ebola virus from the one at large in West Africa, and therefore it must have resulted from an independent spillover. The death toll had risen to 32 by September 6, when the DRC Minister of Health told a press conference that the outbreak can be contained. Probably it can. The city of Kinshasa and its N'Djili International Airport are five hundred miles away, and that's a long bus ride for a mortally ill person.

What will happen next? Nobody knows. That's the best wisdom of science and public health and all other expert prognostication at this point: Nobody knows the future. It's contingent on the scope and the speed of coordinated response, and on luck. What we do know is that the problem of Ebola virus is both acute and chronic. Acute: The West African epidemic of 2014 *must* be contained, and ended, and to do that will require radically more international commitment—in money and material and logistical help and expertise and courageous health workers volunteering to serve in the most difficult conditions—than has been offered so far. Chronic: When that epidemic has been stopped, and the Congolese outbreak too, Ebola virus will not be gone. It will only be hiding again. It will recede into its reservoir host, somewhere amid the forest, and await its next opportunity. We live on a complicated planet, rich with organisms of vast diversity, including viruses, all interacting opportunistically, and although there are seven billion of us humans, the place has not been arranged for our convenience and pleasure.

SOURCE NOTES

8. *"The chimpanzee seems to have been the index case"*: Georges et al. (1999), S70.

24. *"Only limited ecological investigations"*: Johnson et al. (1978), 272.

24. *"No more dramatic or potentially explosive epidemic"*: Ibid., 288.

25. *"No evidence of Ebola virus infection"*: Breman et al. (1999), S139.

30. *"Contact with nature is intimate"*: Heyman et al. (1980), 372–3.

38. *"Viruses of each species have genomes that"*: Towner et al. (2008), 1.

41. *"bad human-like spirits that cause illness"*: Hewlett and Hewlett (2008), 6.

42. *a final "love touch" of the deceased*: Hewlett and Amola (2003), 1245.

44. *"This illness is killing everyone"*: Hewlett and Hewlett (2008), 75.

44. *"Sorcery does not kill without reason"*: Ibid., 75.

46. *"jumped from bed to bed, killing patients left and right"*: Preston (1994), 68.

46. *"transforms virtually every part of the body"*: Ibid., 72.

46. *"suddenly deteriorates," its internal organs deliquescing*: Ibid., 75.

47. *"essentially melts down with Marburg"*: Ibid., 293.

47. *comatose, motionless, and "bleeding out"*: Ibid., 184.

47. *"Droplets of blood stand out on the eyelids "*: Ibid., 73.

49. *even lead to "intestinal destruction"*: Kuhn (2008), 108.

53. *"It is difficult to describe working with a horse infected with Ebola"*: *Yaderny Kontrol* (Nuclear Control) Digest No. 11, Center for Policy Studies in Russia, Summer 1999.

74. *"Taken together, our results clearly point "*: Walsh et al. (2005), 1950.

75. *"Thus, Ebola outbreaks probably do not occur as"*: Leroy et al. (2004), 390.

79. *"the revenge of the rainforest"*: Preston (1994), 289.

86. *Do bats have a different "set point"*: Calisher et al. (2006), 536.

86. *"Emphasis, sometimes complete emphasis, on nucleotide sequence"*: Ibid., 541.

87. *"we are simply waiting for the next"*: Ibid., 540.

87. *"The natural reservoir hosts have not yet been identified"*: Ibid., 539.

92. *"is only one of many such cave populations"*: Towner et al. (2009), 2.

99. *"Patient C was the father of a four-year-old girl"*: Leroy et al. (2009), 5.

100. *"Thus, virus transmission may have occurred"*: Ibid., 6.

101. *"In fact, it is highly likely that several other persons"*: Ibid., 5.

109. *"Within days of his being diagnosed"*: Branswell (2014), 2.

100. *the Gire study found "a rapid accumulation"*: Gire et al. (2014), 1.

100. *"suggests that continued progression of this epidemic"*: Ibid, 2.

SELECT BIBLIOGRAPHY

Baize, Sylvain, Delphine Pannetier, Lisa Oesterich, Tony Rieger, Lamine Koivogui, N'Faly Magassouba, Barré Soropogui et al. 2014. "Emergence of Zaire Ebola Virus Disease in Guinea—Preliminary Report." *The New England Journal of Medicine*, April 16, 2014, online.

Bermejo, Magdalena, José Domingo Rodríguez-Teijeiro, Germán Illera, Alex Barroso, Carles Vilà, and Peter D. Walsh. 2006. "Ebola Outbreak Killed 5000 Gorillas." *Science*, 314.

Biek, Roman, Peter D. Walsh, Eric M. Leroy, and Leslie A. Real. 2006. "Recent Common Ancestry of Ebola Zaire Virus Found in a Bat Reservoir." *PLoS Pathogens*, 2 (10).

Branswell, Helen. "Ebola: See How It Spreads." *Global News*, September 4, 2014.

Breman, Joel G., Karl M. Johnson, Guido van der Groen, C. Brian Robbins, Mark V. Szczeniowski, Kalisa Ruti, Patrician A. Webb, et al. 1999. "A Search for Ebola Virus in Animals in the Democratic Republic of the Congo and Cameroon: Ecologic, Virologic, and Serologic Surveys, 1979–1980." In *Ebola: The Virus and the Disease*, ed. C. J. Peters and J. W. LeDuc. Special issue of *The Journal of Infectious Diseases*, 179 (S1).

Burke, Donald S. 1998. "Evolvability of Emerging Viruses." In *Pathology of Emerging Infections 2*, ed. A. M. Nelson and C. Robert Horsburgh, Jr. Washington: ASM Press

Caillaud, D., F. Levréro, R. Cristescu, S. Gatti, M. Dewas, M. Douadi, A. Gautier-Hion, et al. 2006. "Gorilla Susceptibility to Ebola Virus: The Cost of Sociality." *Current Biology*, 16 (13).

Calisher, Charles H., James E. Childs, Hume E. Field, Kathryn V. Holmes, and Tony Schountz. 2006. "Bats: Important Reservoir Hosts of Emerging Viruses." *Clinical Microbiology Reviews*, 19 (3).

Daszak, Peter, Andrew A. Cunningham, and Alex D. Hyatt. 2000.

"Emerging Infectious Diseases of Wildlife–Threats to Biodiversity and Human Health." *Science's Compass*, 287.

Daszak, P., A. A. Cunningham, and A. D. Hyatt. 2001. "Anthropogenic Environmental Change and the Emergence of Infectious Diseases in Wildlife." *Acta Tropica*, 78.

Dobson, Andrew P., and E. Robin Carper. 1996. "Infectious Diseases and Human Population History." *BioScience*, 46 (2).

Emond, R.T., B. Evans, E. T. Bowen, and G. Lloyd. 1977. "A Case of Ebola Virus Infection." *British Medical Journal*, 2.

Epstein, Jonathan H., Vibhu Prakash, Craig S. Smith, Peter Daszak, Amanda B. McLaughlin, Greer Meehan, Hume E. Field, and Andrew A. Cunningham. 2008. "*Henipavirus* Infection in Fruit Bats (*Pteropus giganteus*), India." *Emerging Infectious Diseases*, 14 (8).

Ewald, Paul W. 1994. *Evolution of Infectious Disease*. Oxford: Oxford University Press.

Formenty, P., C. Boesch, M. Wyers, C. Steiner, F. Donati, F. Dind, F. Walker, and B. Le Guenno. 1999. "Ebola Virus Outbreak among Wild Chimpanzees Living in a Rain Forest of Côte d'Ivoire." In *Ebola: The Virus and the Disease*, ed. C. J. Peters and J. W. LeDuc. Special issue of *The Journal of Infectious Diseases*, 179 (S1).

Garrett, Laurie. 1994. *The Coming Plague: Newly Emerging Diseases in a World Out of Balance*. New York: Farrar, Straus and Giroux.

Georges, A. J., E. M. Leroy, A. A. Renaut, C. T. Benissan, R. J. Nabias, M. T. Ngoc, P. I. Obiang, et al. 1999. "Ebola Hemorrhagic Fever Outbreaks in Gabon, 1994–1997: Epidemiologic and Health Control Issues." In *Ebola: The Virus and the Disease*, ed. C. J. Peters and J. W. LeDuc. Special issue of *The Journal of Infectious Diseases*, 179 (S1).

Gire, Stephen K., Augustine Goba, Kristian G. Andersen, Rachel S. G. Seafon, Daniel J. Park, Lansana Kanneh, Sibirie Jalloh et al. 2014. "Genomic Surveillance Elucidates Ebola Virus Origin and Transmission during the 2014 Outbreak." *Science* Express, August 28, 2014.

Haydon, D. T., S. Cleaveland, L. H. Taylor, and M. K. Laurenson. 2002. "Identifying Reservoirs of Infection: A Conceptual and Practical Challenge." *Emerging Infectious Diseases*, 8 (12).

Hennessey, A. Bennett, and Jessica Rogers. 2008. "A Study of the

Bushmeat Trade in Ouesso, Republic of Congo." *Conservation and Society*, 6 (2).

Hewlett, Barry S., and Richard P. Amola. 2003. "Cultural Contexts of Ebola in Northern Uganda." *Emerging Infectious Diseases*, 9 (10).

Hewlett, Barry S., and Bonnie L. Hewlett. 2008. *Ebola, Culture, and Politics: The Anthropology of an Emerging Disease*. Belmont, CA: Thomson Wadsworth.

Hewlett, B. S., A. Epelboin, B. L. Hewlett, and P. Formenty. 2005. "Medical Anthropology and Ebola in Congo: Cultural Models and Humanistic Care." *Bulletin de la Société Pathologie Exotique*, 98 (3).

Heymann, D. L., J. S. Weisfeld, P. A. Webb, K. M. Johnson, T. Cairns, and H. Berquist. 1980. "Ebola Hemorrhagic Fever: Tandala, Zaire, 1977–1978." *Journal of Infectious Diseases*, 142 (3).

Huijbregts, Bas, Pawel De Wachter, Louis Sosthene Ndong Obiang, and Marc Ella Akou. 2003. "Ebola and the Decline of Gorilla *Gorilla gorilla* and Chimpanzee *Pan troglodytes* Populations in Minkebe Forest, North-eastern Gabon." *Oryx*, 37 (4).

Johnson, Karl M., and Members of the International Commission. 1978. "Ebola Haemorrhagic Fever in Zaire, 1976." *Bulletin of the World Health Organization*, 56.

Johnson, K. M. 1999. "Gleanings from the Harvest: Suggestions for Priority Actions against Ebola Virus Epidemics." In *Ebola: The Virus and the Disease*, ed. C. J. Peters and J. W. LeDuc. Special issue of *The Journal of Infectious Diseases*, 179 (S1).

Karesh, William B. 1999. *Appointment at the Ends of the World: Memoirs of a Wildlife Veterinarian*. New York: Warner Books.

Karesh, William B., and Robert A. Cook. 2005. "The Animal-Human Link." *Foreign Affairs*, 84 (4).

Kuhn, Jens. 2008. *Filoviruses: A Compendium of 40 Years of Epidemiological, Clinical, and Laboratory Studies*. C. H. Calisher, ed. New York: Springer-Verlag.

Lahm, S. A., M. Kobila, R. Swanepoel, and R. F. Barnes. 2006. "Morbidity and Mortality of Wild Animals in Relation to Outbreaks of Ebola Haemorrhagic Fever in Gabon, 1994–2003." *Transactions of the Royal Society of Tropical Medicine and Hygiene*, 101 (1).

Le Guenno, B., P. Formenty, M. Wyers, P. Gounon, F. Walker, and C.

Boesch. 1995. "Isolation and Partial Characterisation of a New Strain of Ebola." *The Lancet,* 345 (8960).

Leirs, Herwig, et al. 1999. "Search for the Ebola Virus Reservoir in Kikwit, Democratic Republic of the Congo: Reflections on a Vertebrate Collection." In *Ebola: The Virus and the Disease,* ed. C. J. Peters and J. W. LeDuc. Special issue of *The Journal of Infectious Diseases,* 179 (S1). Peters and LeDuc (1999).

Leroy, Eric M., Pierre Rouquet, Pierre Formenty, Sandrine Souquière, Annelisa Kilbourne, Jean-Marc Froment, Magdalena Bermejo, et al. 2004. "Multiple Ebola Virus Transmission Events and Rapid Decline of Central African Wildlife." *Science,* 303.

Leroy, Eric M., Brice Kumulungui, Xavier Pourrut, Pierre Rouquet, Alexandre Hassanin, Philippe Yaba, André Délicat, et. al. 2005. "Fruit Bats as Reservoirs of Ebola Virus." *Nature,* 438.

Leroy, E. M., A. Epelboin, V. Mondonge, X. Pourrut, J. P. Gonzalez, J. J. Muyembe-Tamfun, P. Formenty, et al. 2009. "Human Ebola Outbreak Resulting from Direct Exposure to Fruit Bats in Luebo, Democratic Republic of Congo, 2007." *Vector-Borne and Zoonotic Diseases,* 9 (6).

McCormick, Joseph B., and Susan Fisher-Hoch. 1996. *Level 4: Virus Hunters of the CDC.* With Leslie Alan Horvitz. Atlanta: Turner Publishing.

McNeill, William H. 1976. *Plagues and Peoples.* New York: Anchor Books.

Miranda, M. E. 1999. "Epidemiology of Ebola (Subtype Reston) Virus in the Philippines, 1996." In *Ebola: The Virus and the Disease,* ed. C. J. Peters and J. W. LeDuc. Special issue of *The Journal of Infectious Diseases,* 179 (S1).

Monath, Thomas P. 1999. "Ecology of Marburg and Ebola Viruses: Speculations and Directions for Future Research." In *Ebola: The Virus and the Disease,* ed. C. J. Peters and J. W. LeDuc. Special issue of *The Journal of Infectious Diseases,* 179 (S1).

Morse, Stephen S., ed. 1993. *Emerging Viruses.* New York: Oxford University Press.

Nathanson, Neal, and Rafi Ahmed. 2007. *Viral Pathogenesis and Immunity.* London: Elsevier.

Oldstone, Michael B. A. 1998. *Viruses, Plagues, and History.* New York: Oxford University Press.

Pattyn, S. R., ed. 1978. *Ebola Virus Haemorrhagic Fever.* Proceedings of an International Colloquium on Ebola Virus Infection and Other

Haemorrhagic Fevers held in Antwerp, Belgium, December 6–8, 1977. Amsterdam: Elsevier/North-Holland Biomedical Press.

Peters, C. J., and James W. LeDuc, eds. 1999. *Ebola: The Virus and the Disease.* Special issue of *The Journal of Infectious Diseases,* 179 (S1).

Peters, C. J., and Mark Olshaker. 1997. *Virus Hunter: Thirty Years of Battling Hot Viruses Around the World.* New York: Anchor Books.

Pourrut, X., B. Kumulungui, T. Wittmann, G. Moussavou, A. Délicat, P. Yaba, D. Nkoghe, et al. 2005. "The Natural History of Ebola Virus in Africa." *Microbes and Infection,* 7.

Preston, Richard. 1994. *The Hot Zone.* New York: Random House.

Price-Smith, Andrew T. 2009. *Contagion and Chaos: Disease, Ecology, and National Security in the Era of Globalization.* Cambridge, MA: The MIT Press.

Simpson, D.I.H., and the Members of the WHO/International Study Team. 1978. "Ebola Haemorrhagic Fever in Sudan, 1976." *Bulletin of the World Health Organization,* 56 (2).

Sureau, Pierre H. 1989. "Firsthand Clinical Observations of Hemorrhagic Manifestations in Ebola Hemorrhagic Fever in Zaire." *Reviews of Infectious Diseases,* 11 (S4).

Towner, Jonathan S., Tara K. Sealy, Marina L. Khristova, César G. Albariño, Sean Conlan, Serena A. Reeder, Phenix-Lan Quan, et al. 2008. "Newly Discovered Ebola Virus Associated with Hemorrhagic Fever Outbreak in Uganda." *PLoS Pathogens,* 4 (11).

Towner, Jonathan S., Brian S. Amman, Tara K. Sealy, Serena A. Reeder Carroll, James A. Comer, Alan Kemp, Robert Swanepoel, et al. 2009. "Isolation of Genetically Diverse Marburg Viruses from Egyptian Fruit Bats." *PLoS Pathogens,* 5 (7).

Tutin, C.E.G., and M. Fernandez. 1984. "Nationwide Census of Gorilla (*Gorilla g. gorilla*) and Chimpanzee (*Pan t. troglodytes*) Populations in Gabon." *American Journal of Primatology,* 6.

Walsh, Peter D., Roman Biek, and Leslie A. Real. 2005. "Wave-Like Spread of Ebola Zaire." *PLoS Biology,* 3 (11).

Walsh, Peter D., Thomas Breuer, Crickette Sanz, David Morgan, and Diane Doran-Sheehy. 2007. "Potential for Ebola Transmission Between Gorilla and Chimpanzee Social Groups." *The American Naturalist,* 169 (5).

Wamala, Joseph F., Luswa Lukwago, Mugagga Malimbo, Patrick Nguku,

Zabulon Yoti, Monica Musenero, Jackson Amone, et al. 2010. "Ebola Hemorrhagic Fever Associated with Novel Virus Strain, Uganda, 2007–2008." *Emerging Infectious Diseases,* 16 (7).

Weiss, Robin A. 2001. "The Leeuwenhoek Lecture 2001. Animal Origins of Human Infectious Disease." *Philosophical Transactions of the Royal Society of London,* B, 356.

Wolfe, Nathan D., Claire Panosian Dunavan, and Jared Diamond. 2004. "Origins of Major Human Infectious Diseases." *Nature,* 447.

Wolfe, Nathan. 2011. *The Viral Storm: The Dawn of a New Pandemic Age.* New York: Times Books/Henry Holt.

Woolhouse, Mark E. J. 2002. "Population Biology of Emerging and Re-emerging Pathogens. *Trends in Microbiology,* 10 (10, Suppl.).

Zimmer, Carl. 2011. *A Planet of Viruses.* Chicago: The University of Chicago Press.

About the Author

David Quammen is an author and journalist who travels widely to some of the remotest corners of the earth. He writes for a broad range of publications such as *Harper's, Esquire, The Atlantic, Rolling Stone* and the *New York Times*, and is a Contributing Writer at *National Geographic*. His journalism has won him three National Magazine Awards, and he is the recipient of the Academy Award in Literature from the American Academy of Arts and Letters.

Quammen is the author of several acclaimed science and natural history titles, as well as a number of novels. His most recent book, *Spillover*, from which this book is largely extracted, is an exploration into how some of the world's most deadly viruses crossed over from non-human animals into humans. *Spillover* won the Science and Society Book Prize, from the National Association of Science Writers in the United States, and the Society of Biology Book Award in the United Kingdom.

Médecins sans Frontières (Doctors without Borders) and
a few other organizations are providing crucial health care and
support, in complement to national and local health authorities,
within the countries afflicted by the 2014 West African outbreak.
To learn more about MSF's role or to make a donation, go to:
http://www.doctorswithoutborders.org/our-work/medical-issues/ebola